复旦新闻与传播学译库·新媒体系列

吴信训 何道宽 主编

# 社交媒体
## 原理与应用

Social Media
Principles and Applications

[美] 帕维卡·谢尔顿（Pavica Sheldon）著

张振维 译

复旦大学出版社

本书系教育部产学合作协同育人项目(项目编号:201702098003)"大学生社交媒体自我表露行为研究"的研究成果。

# 致　谢

谨以此书献给我的丈夫卢克(Luke)。是他鼓励我写这本书,并在我最需要的时候,不断地给我莫大的支持与爱。

# 目录

- 译者序 ………………………………………… 001
- 前言 …………………………………………… 001

- **第一部分　社交媒体的原理** ………… 001
  - 1　社交媒体与传统人际传播理论 ……… 003
  - 2　社交媒体与大众传播理论 …………… 028
  - 3　社交媒体心理学 ……………………… 049

- **第二部分　社交媒体的应用** ………… 077
  - 4　社交媒体与政治 ……………………… 079
  - 5　社交媒体隐私与安全 ………………… 102
  - 6　社交媒体与教育 ……………………… 114
  - 7　社交媒体与灾难传播 ………………… 135
  - 8　社交媒体与广告 ……………………… 154
  - 9　社交媒体成瘾 ………………………… 167

- **附录　社交媒体简史** ………………… 177

# 译者序

过去这些年,我一直在美国从事社交媒体相关的研究,这个新兴领域近年来在欧美蓬勃发展,不仅在学界受到重视,还与业界之间的联系非常紧密。自2016年我在复旦大学新闻学院从事教学工作起,便深感中国内地的新媒体及社交媒体的发展速度之快、普及度之高,然而中国内地研究社交媒体的相关书籍却十分有限,所以我决定翻译这本书来补充这一领域的文献的不足。

本书中的研究是基于美国的社会情境展开的,而中国内地社交媒体的用户、平台及文化语境都与美国有所不同。本书通过对美国的相关研究进行整理总结,为读者提供一个了解美国社交媒体的平台,为中国化的社交媒体研究与实践提供参考与借鉴。

这本书得以完成,需要感谢许多人,特别感谢复旦大学新闻学院及复旦大学出版社的协助与支持。感谢协助翻译工作的多位学生,他们分别是:邵子瑜,翻译前言、第八章、第九章与附录;刘夕铭,翻译第一章与第二章;陈怡蓓,翻译第三章、第六章与第七章;李沁,翻译第四章与第五章;校稿由邵子瑜、舒瑾涵完成。

张振维<br>
2017年10月于复旦大学新闻学院

# 前　言

　　本书从人际关系、大众媒介、教育、组织和政治环境这些角度探讨了社交媒体的原理与应用。社交媒体技术有许多不同的形式，包括社交网站（如 Facebook 和 Twitter）、博客、维基、在线视频和照片共享网站（如 Pinterest）、点评网站与社交书签网站及视频／文字聊天网站（如 Skype）。前三个章节关注社交媒体的原理。第一章探讨如何将人际传播理论应用于我们对社交媒体的理解中。这一章阐述的理论有：不确定性减少理论、社会渗透理论、社会交换理论、期望违背理论和传播隐私管理理论。第二章讨论了如何将大众传播理论应用于我们对社交媒体的理解中。在这一章中阐释的理论包括使用与满足理论、议程设置理论、框架理论、涵化理论和沉默的螺旋理论。第三章，也是第一部分的最后一章，重点介绍了社交媒体用户的人格心理学和个体差异。这一章概述了媒体心理学研究与社交媒体用户和非用户之间的关系，讨论了与在线自我呈现行为相关的人格特质，包括自恋、外向、自我效能感、对归属的需求及对人气的需求。在社交媒体方面，研究者们关注最多的是自恋与外向倾向，因为这与社交媒体的使用直接相关。其他与社交媒体相关的研究还包括羞怯、孤独和感觉寻求。

　　第二部分共六章，介绍了社交媒体在各种情境下的应用。第四章阐释了社交媒体在政治中的应用，提到了通过 Facebook 和 Twitter 推动的美国与世界各地的政治运动。第五章根据技术在我们的生活中扮演的角色解释了不断变化的隐私定义。第六章探讨了在教育领域中应用社交媒体的益处与挑战,报告了支持在大学课堂中使用博客、YouTube 和 Twitter 的实验研究结果。这一章还讨论了 Facebook 上学生与老师关系的动态。第七章详细介绍了社交媒体在灾难传播中发挥的作用。第八章概述了社交媒体网站中广告的利弊,并进一步提供了如何在 Facebook、Twitter、YouTube、Pinterest 和 LinkedIn 上发布广告的指导方针。最后一章,即第九章,着重于社交媒体成瘾、讨论了有关社交媒体成瘾的定义、原因与后果的问题。尽管很多报纸文章都集中关注了社交媒体消极的方面,但总体而言这方面的研究还是非常少的。

总体来说,本书回答了以下一些问题:谁使用社交媒体?我们能通过社交媒体发展有意义的关系吗?人们在自然灾害和危机发生时是如何使用社交媒体的?人们是如何利用社交媒体来获得社会支持、引发政治变革?隐私已经不存在了吗?雇主是如何使用社交媒体来检查员工的?教师和学生如何使用社交媒体来沟通交流?广告商如何使用社交媒体来宣传他们的产品与服务?为什么人们会沉迷于社交媒体?

第一部分

# 社交媒体的原理

# 1

# 社交媒体与传统人际传播理论

本章主要探讨人际传播理论视域中的社交媒体、主流社交网站和博客。所介绍的理论包括不确定性减少理论（uncertainty reduction）、社会渗透理论（social penetration）、社会交换理论（social exchange）、期望违背理论（expectancy violations）和传播隐私管理理论（communication privacy management）。

## 不确定性减少理论

不确定性减少理论于1975年由贝格尔和卡拉布雷斯（Berger & Calabrese）提出，用于预测和解释陌生人之间的关系发展。根据该理论，随着人们之间感到的不确定性的减少，人际关系会得到发展（Berger，1979；Berger & Calabrese，1975）。大多数人会在不确定的状态中感到不适，并且试图提高对他人行为（行为不确定性）、态度和信任（认知不确定性）的可预测性。如果个体之间无法相互了解，那么他们发展持久关系的可能性就会降低（Berger & Calabrese，1975）。

为减少人际关系中的不确定性，人们会采取各种策略寻求信息（Berger, Gardner, Parks, Schulman, & Miller, 1976）：（1）被动策略：信息寻求者通过观察目标人物的行为来搜集信息；（2）积极策略：这一策略包含为获取他人信息的主动付出，通常是信息寻求者向第三方咨询关于目标人物的信息。（3）互动策略：这一策略需要人们以寻求信息为目的进行直接沟通。一种或多种形式的人际沟通，是减少不确定性的主要方式。随着不确定性程度的降低，信息寻求行为也会减少（不确定性减少理论公理三）。

个体通过面对面互动来降低不确定性的策略，可以被应用于社交网站（social network sites, SNSs）等网络传播语境中。例如，安特尼、瓦尔肯堡和彼得（Antheunis, Valkenburg, & Peter, 2010）检验了社交网站用户为获取他们近期遇到的对象的信息所使用的减少不确定性的策略（被动、积极、互动）。他们发现，Hyves（一个类似Facebook的荷兰网站）的用户使用被动策略最普遍。然而，互动策略是唯一能够有效减少不确定性的策略。其他学者（Parks & Floyd, 1996; Ramirez, Walther, Burgoon, & Sunnafrank, 2002; Tidwell & Walther, 2002）同样指出，在二元计算机中介传播中，参与者主要采用互动策略。

社交网站的用户可以采用很多其他的策略来减少不确定性。例如，他们可以悄悄地查看他人的个人资料（Walther, Van Der Heide, Kim, Westerman, & Tong, 2008），或者向第三方了解他们刚刚在Facebook上加为"好友"的人。尽管我们不能亲眼见到Facebook上的好友，但可以通过他们选择性发布的状态、消息和照片及发布在资料上的个人信息来观察他们的行为（Sheldon & Pecchioni, 2014）。

一个被信息寻求者生动运用的减少不确定性的互动策略是自我表露（self-disclosure）（Berger et al., 1976）。自我表露指有意地分享自己的信息,包括个人经历、想法和态度、感受和价值观,甚至梦想、期待、抱负和目标。惠利斯和格罗茨(Wheeless & Grotz, 1976, p.47)把自我表露定义为"一个人向另一个人传播任何关于自己的信息"。不同于其他形式的计算机中介传播,社交网站通常鼓励用户大量表露关于自己的信息(Antheunis et al., 2010),包括爱好、对音乐书籍和电影的品位、感情状态、性取向等隐私信息(Gross & Acquisti, 2005)。

朱拉德(Jourard, 1971)将在已有关系中产生的表露倾向解释为"二元效应"(dyadic effect)——一个人接收到的信息越多,他进行自我表露的意愿就会越强。一些研究(例如,Craig & Wright, 2012;Sheldon, 2013)发现,一个人对某个Facebook好友所进行的自我表露程度越深、越广,会导致其对该好友的行为有更强的预测性。换言之,朋友之间的相互沟通越多,他们所感受到的不确定性就越少。这些证据支撑了不确定性减少理论中的公理一。公理一提出,随着陌生人之间话语沟通的增多,双方都会感到(对关系中另一方的)不确定性程度降低(West & Turner, 2010)。之后,贝格尔(Berger, 1987)修正了他的理论,认为不确定性减少的过程主要与成熟的关系和初始互动相关。然而,从历史视角来看,不确定性减少理论已经被用于解释传统的面对面互动,近来一些研究表明这一理论也可以用来解释社交网站关系的发展。

## 自我表露与社会吸引力

早期关于关系发展的研究(如Worthy, Gary, & Kahn, 1969)表明,自我表露对于接收者是有益的,而且人们会给那些他们喜欢的人更多的正面奖励。换言之,人们倾向于向他们喜欢的人表露亲密的信息，向不喜欢的人隐瞒亲密的信息（不确定性减少理论公理七，Berger & Calabrese，1975）。这在面对面关系（如 Certner，1973；Fitzgerald，1963；Worthy et al.，1969）和线上传播（Collins & Miller，1994；Levine，2000；Park，Lee，& Kim，2006；Ramirez，Walther，Burgoon，& Sunnafrank，2002）中都被证实是正确的。谢尔顿（Sheldon，2013）讨论了如果 Facebook 好友在相互添加对方之前在社交上已彼此吸引，在添加好友后会如何向对方更多表露自己。这些发现也符合不确定性减少理论定理五（Berger & Calabrese，1975），即认为人们倾向于向他们喜欢的人表露亲密的信息，对不喜欢的人则反之。此外，自我表露可能会激发反向的社会吸引力（Sheldon & Pecchioni，2014）。我们将自己的故事告诉我们喜欢的人，但也倾向于喜欢那些向我们进行自我表露的对象。正如朱拉德（1959）在半个世纪前所提到的，自我表露的举动是个人的奖励和宣泄方式，这种积极的感受导向喜欢。

谢尔顿和佩基奥尼（Sheldon & Pecchioni，2014）发现，就社会吸引、自我表露、行为可预测性的关系而言，无论是纯线上的 Facebook 关系还是纯线下的面对面关系，两者的关系维系过程有着较大的相似性。他们的发现表明，在两个纯 Facebook 朋友之间，社会吸引力和自我表露呈现显著正相关。他们还发现，在纯 Facebook 关系中，自我表露和行为可预测性呈现显著正相关。无论参与者进行互动的媒介是什么，自我表露都与对朋友行为的预测显著相关。

# 自我表露和信任

影响关系中不确定性的另一个因素是信任。

尽管个体之间相互信任的程度很不稳定，但要理解我们何时会选择和别人分享个人信息、何时选择保密,信任在其中发挥着至关重要的作用(Joinson, Reips, Buchanan, & Paine-Schofield, 2011; Kerr, Stattin, & Trost, 1999; Wheeless & Grotz, 1977)。高度信任是亲密的人际关系的标识(Anderson & Emmers-Sommer, 2006; Bukowski & Sippola, 1996; Rempel, Holmes, & Zanna, 1985)。信任对于减少面对面交流的不确定性十分重要(Dainton & Aylor, 2001),应用于社交网站也是如此。在一项检验最好的Facebook关系的研究中,谢尔顿(2009)发现,我们对另一个人的行为越确定,我们对他/她的信任度就越高。谢尔顿和佩基奥尼(2014)发现,我们对我们纯Facebook好友及纯面对面好友越信任,我们就会向他们进行更多的自我表露。我们越信任他们,我们就越能预测他们的行为。这些研究符合早期面对面研究(Foubert & Sholley, 1996; Steel, 1991)的结果,研究表明在所有的性格因素中,信任在预测自我表露中影响力最大。然而,当对比两对朋友之间的信任时,谢尔顿和佩基奥尼(2014)的研究发现,被试者对他们Facebook好友的信任要少于对面对面好友的信任。贝恩、科尼什和艾斯帕梅·坎普曼(Bane, Cornish, & Erspamer Kampman, 2010)检验了女性博主对线上与真实生活中同性友谊的看法,他们发现相比线上友谊,参与者会感到更容易在真实生活的友谊中产生信任、忠诚、感情支持和实际帮助。这些发现支持了既往研究(Parks & Roberts, 1998),即表明线上关系在相互依赖、理解和承诺上都弱于现实友谊。

## 社会渗透理论

另一个有助于解释自我表露在关系发展中的作用的理论是社

会渗透理论。奥尔特曼和泰勒（Altman & Taylor，1973）对社会渗透理论进行概念化，来证实关系结合（社会渗透）的过程会使关系从表面发展到亲密。亲密可以从身体的、智力的和情感的这三方面来定义，而社会渗透的过程则包含言语行为和非言语行为。大部分关系在走向亲密的过程中都会遵循一些特定的轨迹或路径。根据相关理论，关系发展是一个有体系、可预测、循序渐进的过程，而自我表露则是其核心。奥尔特曼和泰勒借洋葱来解释自我表露，认为表露首先从外层的表皮开始，逐渐到洋葱的核心。一个人的外层是那些可以向他人展示的东西，其中所包含的一些关于自身的表面信息（如对音乐、衣服、食物等的偏好）可以在关系的早期阶段与他人分享。动机充足的信息寻求者，则会试图渗透每一层，直到抵达另一个人自我的核心。洋葱内在的核心往往是那些仅被少数人知道的信息，包括强烈的感情、观念、信仰和自我意识（Altman & Taylor，1973）。

这种渗透可以从两个维度进行观察：广度和深度。广度是指在关系中讨论的不同类型话题的数量，深度是指引导话题讨论的亲密程度。有深度的信息对我们的自我认同更为重要。奥尔特曼和泰勒（1987）认为，随着关系走向亲密，更多的话题会被讨论（广度），其中一些话题会被深入探讨（深度）。

有研究表明，人们在互联网关系中的表露显著多于现实关系（Parks & Floyd，1996），而一些近期的研究（Tang & Wang，2012；Sheldon & Pecchioni，2014）则挑战了这些观点。唐和王（Tang & Wang，2012）对1 027名台湾博主进行了问卷调查，探究这些博主在其博客上进行自我表露的话题，以及博主面对三种目标受众（线上读者、挚友、父母）时自我表露的广度和深度。唐和王发现，在关于想法、感受和经历的自我表露上，这些博主会对他们现实生活中的挚友进行最深、最广的表露,而非对他们的父母或线上读者。他们似乎能意识到线上自我表露的风险,从而避免透露个人及经济情况。谢尔顿和佩基奥尼(2014)也指出,大学生对他们 Facebook 好友的自我表露少于现实好友。然而,无论是在 Facebook 还是在面对面讨论中,社会吸引力都是话题数量和变化性(广度)最重要的预测指标,亲密程度则是谈话深度(自我表露的深度维度)最重要的预测指标(Sheldon & Pecchioni, 2014)。这支持了"洋葱模型"——随着我们对某个人了解的加深,我们会倾向于向对方表露更亲密的话题。而"为什么社会吸引力只对我们和他人谈论的话题数量有影响"这一问题的答案也相应清晰。另有研究(Krasnova, Spiekermann, Koroleva, & Hildebrand, 2010)则认为,自我表露中的深度与社交网站没有关系。自我表露的深度是一个"高度主观的变量",就好像"一个平台的经济价值,不是由活跃用户的表露程度定义的,而是由他们的参与和互动来定义的"(Krasnova et al., 2010, p.113)。

吉恩(Jin, 2013)分析了在 Twitter 上进行自我表露的层次。结果显示,自我表露包含五个部分:(1)日常生活和娱乐;(2)社会认同;(3)能力;(4)社会经济地位;(5)健康。与日常生活和娱乐相关的信息位于自我表露"洋葱模型"的最外层,是用户表露最频繁(自我表露的广度维度)的信息类型。和健康相关的私人信息在"洋葱模型"的最里层,这意味着用户不想将这些信息公开(深度维度)。

社交媒体上的自我表露有积极的影响或效果。奥尔森(Olson, 2012)通过问卷和焦点小组展开研究,询问参与者在 Facebook 上自我表露时的感受如何。81%的问卷参与者同意他们在 Facebook 上自我表露时感受良好,大多数焦点小组的参与者同意在Facebook上进行自我表露对他们的自尊心有积极的影响。他们指出，大部分人会在Facebook上发布正面信息，并因此对自我感觉非常良好。布龙斯坦（Bronstein，2013）对博客的研究做了综述，发现私人博客逐渐变成进行自我表露的主要平台，人们在这里向陌生人表露个人信息——往往更不害怕潜在的批评或嘲笑。大多数博主都倾向于公开真实姓名，因为这并不是个人匿名，而是社交匿名——"身体的不可见性构成了线上沟通的特征，并且赋予博主支配他们博客内容的权力，以及选择向读者表露或隐藏身份的权力"（Bronstein，2013，p.173）。布龙斯坦（2012）指出，在线上环境中，博主能够通过发泄情绪、想法和观点来释放感情压力，这比面对面沟通的风险要低。麦肯齐（Mckenzie，2008）、柯和陈（Ko & Chen，2009）也指出，博主会通过自我表露感受到积极的情绪，如满足和兴奋，这显著影响了他们对幸福的主观认知。

社会渗透理论以社会交换理论的几个原则为基础，将在下文中进行详细叙述。

# 社会交换理论

在1959年，蒂博和凯利（Thibaut & Kelley）写道："只要在回报和付出上能够得到满足，每个个体就会自愿加入并且留在任何关系中。"（p.37）蒂博和凯利的这个理论最初被命名为依赖理论，现在以"社会交换理论"闻名。该理论认为，人们会从付出和回报的角度来评价他们的关系。付出是关系中对人有负面价值的部分（如压力、时间、精力、注意力）。回报是关系中对人有正面价值的部分（如愉快、忠诚、注意力）（引自West & Turner，2010）。萨巴泰利和希恩（Sabatelli & Shehan，1993）用市场来类比关系如何发挥作用。根据社会交换理论（SET），一段关系的价值预测了关系的结果。积极的关系是那些有正面价值的关系（如回报多于付出），消极的关系是那些有负面价值的关系（如付出多于回报）。积极的关系会一直维持下去，而消极的关系更可能会结束（West & Turner, 2010）。

社会交换理论的第一个假设是，关系是相互依赖的。一段关系的结果不会由个人单方面决定，而是由关系中的双方共同创造的。无论何时，关系中的一方有所行动，关系的另一方和整体关系本身都会受到影响。这一理论的第二个假设是，关系的发展是一个过程。时间会影响交换，因为一段关系中过去的经历通常会被用来评价和预测关系中的回报和付出。人们用来评价付出和回报的标准也会因人而异、因时而变。对于某人来说是回报，对于他人来说可能就是付出，反过来也是如此（West & Turner, 2010）。

在蒂博和凯利（1959）看来，评估一段关系的标准还包括：比较水平和选择比较水平。比较水平（comparison level, CL）是指人们所感到的应从一段特殊关系里获得回报和付出的标准。选择比较水平（comparison level for alternatives, CLalt）是指一个人在关系中，愿意接受的回报的最低水平（Roloff, 1981）。一个人通常会以替代关系或独处中可获得的回报来测量他们所愿意接受的回报的最低水平。一些学者（如 Walker, 1984）认为，选择比较水平提供了一个很好的方法来解释为什么一些女性会处于被虐待的关系中。

一些研究把社会交换理论应用到社交网络中。拉塞尔（Drussell, 2012）认为，一个人在 Facebook 上编辑或发送评论、更新状态所用的时间和精力与他所感知的回报直接相关——包括"赞"的数量或回复。更新 Facebook 或 Twitter 需要很少的时间和精力,但潜在的回报是没有限制的——尤其对于那些拥有大量读者的人而言。拉塞尔(2012)讨论了和社会交换理论相关的权力概念。权力是指对回报和惩罚的控制。一些人拥有社会权力,因此拥有影响他人想法和行动的能力。换言之,一个人在社会网络中拥有越多的朋友,这个人就会有越高的地位和权力。这个权力也可能反映在拒绝他人的好友请求上。

卡拉斯诺娃等人(Krasnova et al., 2010)也通过焦点小组的方法,使用社会交换理论研究了用户在 Facebook 和 StudiVZ(一个流行于德国的社交网站)上表露个人信息的动机。结果表明,关系保持的便捷性是引导用户在线上社交网站平台分享信息的最重要的因素。此外,表露个人信息的动机还伴有建立新关系的愉快与渴望。其他研究也证实了维持关系是使用 Facebook 的最主要动机(如 Sheldon, 2008)。从付出的角度看,卡拉斯诺娃等人发现感知到的隐私风险是抑制用户在社交网站平台上表露信息的主要因素。参与者提到,在决定是否进行自我表露时,他们会进行有意识的"隐私计算"。根据社会交换理论,人们主要从付出和回报的角度来评价他们的关系。

在卡拉斯诺娃等人(2010)的研究中,用户承认他们在社交平台上自我表露的时候意识到了风险,但是,为获得某些好处,他们依然选择这样做。这些好处包括:(1)通过发布一条状态,高效地同时和大量朋友沟通;(2)建立友谊——进行更多自我表露,从而建立新的友谊。

## 期望违背理论

另一个用来解释面对面互动,并且也可以用来解释线上关系的理论是期望违背理论（Burgoon，1978）。根据这一理论，人们的互动是由期望驱动的。有三个因素会影响期望：个体传播者因素（性别、性格、年龄、外貌）、关系因素（关系的历史、地位差异、吸引和喜欢的程度）及语境因素（正式/非正式、社会/任务功能、环境限制、文化规范）（Burgoon & Hale，1988）。

人类对行为的期望是后天养成的。人们从他们出生地的文化中习得他们的期望。比如，在美国，人们对于师生关系的期望是，老师拥有很多学科问题上的知识，并且能够给学生提供帮助。当期望被违背时，它既可能被视为积极的，也可能被视为消极的，其评判取决于他人的潜在回报。韦斯特和特纳（West & Turner，2010）为期望违背提供了一个例子：在公交车上，陌生人的持久凝视给人的友好感会远低于恋人的持久凝视。

最初，"期望违背"的概念被用来解释对行为规范的非言语违背。之后逐渐变为同时解释语言和非言语的违背。近来，该概念开始被用来解释社交网站上的行为。比如，麦克劳克林和维塔克（McLaughlin & Vitak，2012）探索了社交网站上的规范是如何随时间而变化以及对这些规范的违背如何影响个体的自我呈现和人际关系的目标。他们假设，Facebook 的规范可能和传统的线下规范不同，因为 Facebook 的友谊包括大范围的关系，涵盖了亲密朋友、大学同学、家庭成员和熟人。Facebook 的规范存在表达的含蓄性——它们并没有被写下来，而是被团体普遍理解（Burnett & Bonnici，2003）。在麦克劳克林和维塔克（2012）的研究中，焦点小组成员表示，他们通过观察其他成员的表现，学会了如何管理自己。因此，如果看到朋友发布更多的信息，用户就也会这么做。几乎所有参与者均同意，如果忽视一个有过私交的朋友的请求，会显得很粗鲁。这就是所谓的期望规范。此外，还有对朋友圈和私人信息的使用规范。参与者表示，他们期待用朋友圈和状态更新去分享视频、玩笑、生日愿望等，而用私信或聊天的形式来透露私人信息、组织活动或开展小范围群体的讨论。Facebook规范违背中被最频繁提及的是过多的状态更新，其次是过于情绪化的状态，包括关系争执和其他公共争执。麦克劳克林和维塔克（2012）指出，对负面期望违背的最常见的回应是删除Facebook的"好友"或者"屏蔽"他们的消息动态。

Facebook上的师生关系则是另一个可能发生期望违背的领域。一些研究曾告诫教职员工在添加学生为Facebook好友的时候要保持警惕。卡尔和佩鲁切特（Karl & Peluchette, 2011）及施瓦茨（Schwartz, 2009）主张，教职员工要对新技术持开放态度，但不可主动与学生开展友谊，仅仅回复他们的请求即可，从而保持被动的姿态。正如卡尔和佩鲁切特（2011）所发现的，一些学生会对来自他们教授的请求表示愤怒。在马勒斯基和彼得斯（Malesky & Peters, 2012）的研究中，近40%的学生和30%的教员认为，教授不适合拥有社交平台账户。只有在教授试图帮助学生（如记住他们的名字、提供额外的加分）时，学生看到教授的行为被视为更合适。

2013年，谢尔顿以教职工为调查对象，从而了解他们对于添加学生为好友的期望。50%的被调查者表示，他们的Facebook拥有一个及以上学生好友。该数字远超2010年美国俄亥俄州立大学药学院关于教职工与学生友谊的调查数据（Metzger, Finley, Ulbrich, & McAuley, 2010）。这可能表明对于非正式的师生关系，人们的态度正发生转变，更多教职工意识到与学生成为Facebook好友的好处。随着态度改变，规范也发生了变化（Sheldon, 2014），这些规范反映了用户和谁分享私人信息。

# 传播隐私管理理论

根据传播隐私管理理论（Petronio，2002），人们会基于以文化、性别、语境等为标准的"心理计算"，对于告诉别人、隐瞒别人的内容作出选择并制定规则（引自 West & Turner，2010）。彼得罗尼奥（Petronio）使用"隐私表露"而非"自我表露"的概念。隐私表露是指传播私人信息给他人的过程。这一理论进一步指出，在公开信息和隐私的转变之间有一条线（隐私界限）。当人们把私人信息表露给别人时，他们所分享信息的界限就被称为集体界限。

人们在决定要表露或隐藏个人信息时，主要依据五个标准：文化、性别、动机、语境和风险收益比。例如，不同文化有不同的关于隐私和公开的规范。男性和女性在社会化过程中会建立不同的关于隐私和公开的操作规则。人们可能被鼓励去表露个人信息，从而发展亲密关系。特定环境也会引发或阻止表露。最后，规则被建立在风险收益的标准之上，这和社会交换理论的规则基础十分相似。在表露个人信息时，人们会估算相比收益而产生的风险（引自 West & Turner，2010）。

根据对传播隐私管理理论的控制权和所有权的推测，人们认为他们拥有自身的隐私信息，并且控制授予访问权限给信任的人。当通向隐私信息的入口关闭，界限就十分厚实。当入口打开，人们便拥有相对单薄的界限。界限扰动包含界限期望和规范的冲突。当信息共有者没有明确协商如何向第三方分享隐私信息时，扰动就可能出现（Petronio，Jones，& Morr，2003）。传播隐私管理是一个辩证的理论，这意味着人们既需要（自治权的）保护，又需要分享(社会互动)。彼得罗尼奥(2002)指出,这两种力量会不断相互影响,因此隐私的张力不能被简化为二元论。

最近,该理论开始被用于解释"隐私悖论",即为什么人们一边在社交网站上表露私人信息,一边又表达出对隐私的担忧。博伊德(boyd,2010)认为,"一场在走廊中进行的对话,会被默认是私密的,有意为之才会变成公共的",但在Facebook留言板上"对话会被默认是公共的,有意为之才会变成隐私"。换言之,面对面分享的信息会被人们轻易忘记,但在线上分享的信息会被储存和归档,并且是可以复制的。吉恩(2013)讨论了在Twitter上被保护的推文(厚边界)(Petronio,2002)和公共推文(薄边界)(Petronio,2002)之间的差异。Twitter用户在个人页面上发布的推文是自我生产的内容(可以是保护推文也可以是公共推文),而朋友圈中的其他的推文则是他人表露的公共推文。吉恩对Twitter的用户、非用户和离开平台的用户进行了调查,发现非用户普遍对隐私有所顾虑,通常不愿注册Twitter账户。

教师会基于日常经验对隐私问题进行辩证处理:他们想要向学生表露什么信息、隐藏什么信息。然而,大多数研究发现教师自我表露和学生对情感学习的认知之间呈正相关(例如,Sorensen,1989;Andersen,Norton,& Nussbaum,1981)。一名教师的言语行为,如使用个人举例,会创造一个更直接的教学环境——这与情感、认知学习也呈正相关(Christensen & Menzel,1998;Gorham,1988)。马泽、墨菲和西蒙兹(Mazer,Murphy,& Simonds,2007)发现,当一名女性教师表露一些特定信息,如私人照片或对特定话题的观点,学生便会感知到他们和教师之间的相似之处。大多数被调查的大学生认为教师使用Facebook是积极的,但他们也指出教授应该在Facebook上"做自己",这样他们能够"对自身的人格有更深的感受"。

蔡尔德、彼得罗尼奥、阿杰曼－布杜和韦斯特曼（Child, Petronio, Agyeman-Budu, & Westermann, 2011）立足于传播隐私管理理论，以356名私人日记类型的博主为调查对象，探讨博客上的隐私规则适应过程，并检查激发博主删除或修改推送（"博客清理"）的状况。调查发现，这些博主往往会被风险评估和删除内容的动机所驱动，而去回顾写过的内容。这种删除信息的渴望被解释为博主对博客内容是否准确表达自我的高度关注。此外，马登和史密斯（Madden & Smith, 2010）指出，47%的年轻成年用户会通过删除评论和推送来清理博客。综上，他们采用了事后隐私管理决定标准（Petronio, 2002）。

梅茨格和普雷（Metzger & Pure, 2009）研究了 Facebook 用户在什么情况下会选择为个人信息建立较薄、适中或较厚的边界。他们从三方面入手：(1) 个人采取的隐私设置；(2) 个人档案中表露的信息量；(3) 个人的"好友添加"行为。研究结果表明，Facebook 用户更倾向于用较薄到适中的边界来平衡对隐私和公开信息之间的矛盾欲望。用户们指出，他们更倾向于表露能够帮他们保持或提高社会关系的信息，较少表露对关系形成和私人安全有负面影响的信息。大多数参与者的好友添加行为非常自由，可以与他们从未见过的人成为好友。从性别差异角度来看，女性会对发布在线上的个人信息有更强烈的控制欲望，她们倾向于发布自己和别人的照片，但较少表露联系信息（Metzger & Pure, 2009; Tufekci, 2008）。

吉布斯和曹（Gibbs & Cho, 2010）以美国的 Facebook 用户和韩国的 Cyworld 用户为研究对象，检验了社交媒体使用者中隐私管理的文化差异。其中，Facebook 用户倾向于控制与他们档案可见性相关的隐私设置，而 Cyworld 用户倾向于控制可检索性。Facebook 用户比 Cyworld 用户有更多好友，但关系亲密度较低。在此，或许可以用个人主义和集体主义的文化观念差异来解释吉布斯和曹（2010）的发现。韩国人对于圈子外的成员设有厚边界，但对圈子内的成员则有薄边界。此前一些研究也表明，集体文化会严格区分圈内和圈外的成员（如 Triandis，1989）。换言之，集体文化的成员是为了与有较少隐私顾虑的朋友保持亲密的圈子，而不是为了获得社会资本才去使用社交网站平台。埃利森、斯坦菲尔德、兰珀（Ellison, Steinfield, & Lampe, 2007）曾发现 Facebook 对于美国学生的主要好处是保持弱关系。

传播隐私管理理论也被用来检验已成年的 Facebook 年轻用户如何看待父母的 Facebook 好友请求。蔡尔德和韦斯特曼（2013）研究了 235 个用户，发现大多数人会接受父母的 Facebook 好友请求，并且并未就这些请求对隐私规则作出任何严格的调整。然而，那些接受母亲的好友请求并且不对隐私规则作出更严格调整的人，通常来自重视公开、透明观念的家庭，这些人也和他们的母亲有着更高质量的关系。然而亲密和信任却不是接受父亲好友请求的重要因素。蔡尔德和韦斯特曼（2013）通过归因到父母在家庭中所扮演的角色来解释这一发现。母亲在孩子的成长过程中扮演了照顾者的角色，比起父亲，和孩子之间有着更亲密的关系。而在青春期后期，子女往往感到父亲扮演着纪律执行者的角色（McKinney & Renk, 2008）。

总而言之，大多数关于面对面环境的自我表露研究，强调通过表露来建立新关系的益处。而大多数关于线上表露的研究则强调表露的风险。

# 总　　结

本章对比了社交媒体对线上传播与线下传播的影响，呈现了五种检验和解释关系发展的理论：不确定性减少理论、社会渗透理论、社会交换理论、期望违背理论和传播隐私管理理论。每种理论均检验了是什么直接或间接地推动参与者与他人的关系从陌生到熟悉。

结论表明，两种传播语境本质上是相似的，即都会采用相同的策略来建立关系，但二者仍有差异性。例如，一个采用不确定性减少理论的研究发现，尽管在预测线上和线下自我表露时，信任都是核心要素，但实际上，被测试者更信任面对面的朋友。本章检验的另一个特殊要素是"隐私悖论"，这一要素仅仅和社交网站上的个人信息表露有关。该悖论的一个基础是，当用户持续在线上表露个人信息时，他们会对表露的反响保持更多警惕。这个悖论在线下沟通中并不存在，因为信息不会被存储和归档以让后人查阅到。

# 参 考 文 献

Altman, I., & Taylor, D. (1973). *Social penetration: The development of interpersonal relationships.* New York: Holt, Rinehart, Winstron.

Altman, I., & Taylor, D. (1987). Communication in interpersonal relationships: Social penetration processes. In M. Roloff & G. Miller (Eds.), *Interpersonal processes* (pp. 257-277). London, England: Sage Publications.

Andersen, J. F., Norton, R. W., & Nussbaum, J. F. (1981). Three investigations exploring relationships between perceived teacher communication behaviors and student learning. *Communication Education*,

30, 377-392. doi: 10.1080/03634528109378493.

Anderson, T. A., & Emmers-Sommer, T. M. (2006). Predictors of relationship satisfaction in online romantic relationships. *Communication Studies*, *57*, 153-172. doi: 10.1080/10510970600666834.

Antheunis, M. L., Valkenburg, P. M., & Peter, J. (2010). Getting acquainted through social network sites: Testing a model of online uncertainty reduction and social attraction. *Computers in Human Behavior*, *26*, 100-109. doi: 10.1016/j.chb.2009.07.005.

Bane, C., Cornish, M., Erspamer, N., & Kampman, L. (2010). Self-disclosure through weblogs and perceptions of online and "real-life" friendships among female bloggers. *CyberPsychology, Behavior & Social Networking*, *13*, 131-139. doi: 10.1089/cyber.2009.0174.

Berger, C. R. (1979). Beyond initial interaction: Uncertainty, understanding, and the development of interpersonal relationships. In H. Giles & R. St. Clair (Eds.), *Language and social psychology* (pp. 122-144). Oxford: Basil Blackwell. doi: 10.1111/j.1468-2958.1975.tb00258.x.

Berger, C. R. (1987). Communicating under uncertainty. In M. E. Roloff & G. R. Miller (Eds.), *Interpersonal processes* (pp. 39-62). Newbury Park, CA: Sage.

Berger, C. R., & Calabrese, R. J. (1975). Some explorations in initial interaction and beyond: Toward a developmental theory of interpersonal communication. *Human Communication Research*, *1*, 99-112. doi: 10.1111/j.1468-2958.1975.tb00258.x.

Berger, C. R., Gardner, R. R., Parks, M. R., Schulman, L. S., & Miller, G. R. (1976). Interpersonal epistemology and interpersonal communication. In G. R. Miller (Ed.), *Explorations in interpersonal communication* (pp. 149-172). Newbury Park, CA: Sage Publications.

boyd, d. m. (2010). Making sense of privacy and publicity. SXSW. Austin, Texas.

Bronstein, J. (2012). Blogging motivations for Latin American blogosphere: A uses and gratifications approach. In T. Dumova & E. Fiordo (Eds.), *Blogging in the global society: Cultural, political and geographical aspects* (pp. 200-215). Hershey, PA: Information Science Reference. doi: 10.

4018/978-1-60960-744-9.ch012.

Bronstein, J. (2013). Personal blogs as online presences on the internet: Exploring self-presentation and self-disclosure in blogging. *Aslib Proceedings*, *65*, 161-181. doi: 10.1108/00012531311313989.

Bukowski, W. M., & Sippola, L. K. (1996). Friendship and morality: (How) are they related? In W. M. Bukowski, A. F. Newcomb, & W. W. Hartup (Eds.), *The company they keep* (pp. 238-261). Cambridge, MA: Cambridge University Press.

Burgoon, J. (1978). A communication model of personal space violations: Explication and an initial test. *Human Communication Research*, *4*, 129-142. doi: 10.1111/j.1468-2958.1978.tb00603.x.

Burgoon, J. K., & Hale, J. L. (1988). Nonverbal expectancy violations: Model elaboration and application to immediacy behaviors. *Communication Monographs*, *55*, 58-79. doi: 10.1080/03637758809376158.

Burnett, G., & Bonnici, L. (2003). Beyond the FAQ: Explicit and implicit norms in Usenet news-groups. *Library and Information Science Research*, *25*, 333-351. doi: 10.1016/S0740-8188(03)00033-1.

Certner, B. C. (1973). Exchange of self-disclosures in same-sexed groups of strangers. *Journal of Consulting and Clinical Psychology*, *40*, 292-297. doi: 10.1037/h0034446.

Child, J. T., Petronio, S., Agyeman-Budu, E. A., & Westermann, D. A. (2011). Blog scrubbing: Exploring triggers that change privacy rules. *Computers in Human Behavior*, *27*, 2017-2027. doi: 10.1016/j.chb.2011.05.009.

Child, J. T., & Westerman, D. A. (2013). Let's be Facebook friends: Exploring parental Facebook friend requests from a communication privacy management (CPM) perspective. *Journal of Family Communication*, *13*, 46-59. doi: 10.1080/15267431.2012.742089.

Christensen, L. J., & Menzel, K. E. (1998). The linear relationship between student reports of teacher immediacy behaviors and perceptions of state motivation, and of cognitive, affective, and behavioral learning. *Communication Education*, *47*, 82-90. doi: 10.1080/03634529809379112.

Collins, N. L., & Miller, L. C. (1994). Self-disclosure and liking: A meta-analytic review. *Psychological Bulletin*, *116*, 457-475. doi: 10.1037/0033-2909.116.3.457.

Craig, E., & Wright, K. B. (2012). Computer-mediated relational development and maintenance on Facebook. *Communication Research Reports*, *29*, 119-129. doi: 10.1080/08824096.2012.667777.

Dainton, M., & Aylor, B. (2001). A relational uncertainty analysis of jealousy, trust, and maintenance in long-distance versus geographically close relationships. *Communication Quarterly*, *49*, 172-188. doi: 10.1080/01463370109385624.

Drussell, J. (2012). *Social networking and interpersonal communication and conflict resolution skills among college freshmen*. Master of Social Work Clinical Research Paper, 21. Retrieved from http://sophia.stkate.edu/msw_papers/21.

Ellison, N., Steinfield, C., & Lampe, C. (2007). The benefit of Facebook "friends: " Social capital and college students' use of online social network sites. *Journal of Computer-Mediated Communication*, *12*(4), article 1. doi: 10.1111/j.1083-6101.2007.00367.x.

Fitzgerald, M. P. (1963). Self-disclosure and expressed self-esteem, social distance, and areas of the self revealed. *The Journal of Psychology*, *56*, 405-412. doi: 10.1080/00223980.1963.9916655.

Foubert, J., & Sholley, B. K. (1996). Effects of gender, gender role, and individualized trust on self-disclosure. *Journal of Social Behavior and Personality*, *11*, 277-288.

Gibbs, J. & Cho, S. E. (2010). *A cross-cultural investigation of privacy management in Facebook and Cyworld*. Presented at the International Communication Association conference.

Gorham, J. (1988). The relationship between verbal teacher immediacy behaviors and student learning. *Communication Education*, *37*, 40-53. doi: 10.1080/03634528809378702.

Gross, R., & Acquisti, A. (2005). Information revelation and privacy in online social networks (The Facebook case). In *ACM workshop on privacy in the electronic society* (pp. 71-80). Alexandria: USA. doi: 10.1145/1102199.1102214.

Jin, S. A. (2013). Peeling back the multiple layers of Twitter's private disclosure onion: The roles of virtual identity discrepancy and personality traits in communication privacy management on Twitter. *New Media & Society*, *15*, 813–833. doi: 10.1177/1461444812471814.

Joinson, A. N., Reips, U. D., Buchanan, T. B., & Paine-Schofield, C. B. (2011). Privacy, trust, and self-disclosure online. *Human-Computer Interaction*, *25*, 1–24. doi: 10.1080/07370020903586662.

Jourard, S. M. (1959). Self-disclosure and other-cathexis. *Journal of Abnormal and Social Psychology*, *59*, 428–431. doi: 10.1037/h0041640.

Jourard, S. M. (1971). *Self-disclosure: An experimental analysis of the transparent self*. New York: Robert E. Krieger.

Karl, K. A., & Peluchette, J. V. (2011). "Friending" professors, parents and bosses: A Facebook connection conundrum. *Journal of Education for Business*, *86*, 214–222. doi: 10.1080/08832323.2010.507638.

Kerr, M., Stattin, H., & Trost, K. (1999). To know you is to trust you: Parents' trust is rooted in child disclosure of information. *Journal of Adolescence*, *22*, 737–752. doi: 10.1006/jado.1999.0266.

Ko, H., & Chen, T. (2009). *Understanding the continuous self-disclosure of bloggers from the cost-benefit perspective*. Proceedings of the 2nd Conference on Human System Interactions. Cantania, Italy. doi: 10.1109/HSI.2009.5091033.

Krasnova, H., Spiekermann, S., Koroleva, K., & Hildebrand, T. (2010). Online social networks: Why we disclose. *Journal of Information Technology*, *25*, 109–125. doi: 10.1057/jit.2010.6.

Levine, D. (2000). Virtual attraction: what rocks your boat. *Cyber Psychology & Behavior*, *3*, 565–573. doi: 10.1089/109493100420179.

Madden, M., & Smith, A. (2010). *Reputation management and social media: How people monitor their identity and search for others online*. Pew Internet and American Life Project website. Retrieved from http://www.pewinternet.org/Reports/2010/Reputation-Management.aspx.

Malesky, L. A., & Peters, C. (2012). Defining appropriate professional behavior for faculty and university students on online social networking websites. *Higher Education*, *63*, 135–151. doi: 10.1007/s10734–011–

9451-x.

Mazer, J. P., Murphy, R. E., & Simonds, C. J. (2007). I'll see you on "Facebook": The effects of computer-mediated teacher self-disclosure on student motivation, affective learning, and classroom climate. *Communication Education*, *56*, 1–17. doi: 10.1080/03634520601009710.

McKenzie, H. M. (2008). *Why bother blogging? Motivations for adults in the United States to maintain a personal journal blog*. Unpublished master's thesis. North Carolina State University, Raleigh, NC.

McKinney, C., & Renk, K. (2008). Differential parenting between mothers and fathers: Implications for late adolescents. *Journal of Family Issues*, *29*, 806–827. doi: 10.1177/0192513X07311222.

McLaughlin, C., & Vitak, J. (2012). Norm evolution and violation on Facebook. *New Media and Society*, *14*, 299–315. doi: 10.1177/1461444811412712.

Metzger, A. H., Finley, K. N., Ulbrich, T. R., & McAuley, J. W. (2010). Pharmacy faculty members' perspectives on the student/faculty relationship in online social networks. *American Journal of Pharmaceutical Education*, *74*(10), 188.

Metzger, M., & Pure, R. (2009). *Privacy management in Facebook*. Presented at the annual meeting of National Communication Association.

Olson, A. M. (2012). *Facebook and social penetration theory*. Master's thesis. Gonzaga University.

Park, J. Y., Lee, J. E., & Kim, N. (2006). *"Hi! My name is Clora": The effects of self-disclosing agents on the attitude and behavior of users*. Presented at the annual meeting of the International Communication Association, Dresden, Germany.

Parks, M. R., & Floyd, K. (1996). Making friends in cyberspace. *Journal of Communication*, *46*, 1–17. doi: 10.1111/j.1460-2466.1996.tb01462.x.

Parks, M. R., & Roberts, L. D. (1998). Making MOOsic: The development of personal relationships on line and a comparison to their off-line counterparts. *Journal of Social and Personal Relationships*, *15*, 517–537. doi: 10.1177/0265407598154005.

Petronio, S. (2002). *Boundaries of privacy: Dialectics of disclosure*. New

York: State University of New York Press.

Petronio, S., Jones, S., & Morr, M. C. (2003). Family privacy dilemmas: Managing communication boundaries within family groups. In L. R. Frey (Ed.), *Group communication in context: Studies of bona fide groups* (pp. 23–55). Mahwah, NJ: Erlbaum.

Ramirez, A., Jr., Walther, J. B., Burgoon, J. K., & Sunnafrank, M. (2002). Information seeking strategies, uncertainty, and computer-mediated communication: Toward a conceptual model. *Human Communication Research, 28*, 213–228. doi: 10.1111/j.1468-2958.2002.tb00804.x.

Rempel, J. K., Holmes, J. G., & Zanna, M. P. (1985). Trust in close relationships. *Journal of Personality and Social Psychology, 49*, 95–112. doi: 10.1037/0022-3514.49.1.95.

Roloff, M. E. (1981). *Interpersonal communication: The social exchange approach.* Beverly Hills, CA: Sage.

Sabatelli, R. M., & Shehan, C. L. (1993). Exchange and resource theories. In P. G. Boss, W. J. Doherty, R. LaRossa, W. R. Schumm, & S. K. Steinmetz (Eds.), *Sourcebook of family theories and methods: A contextual approach* (pp. 385–411). New York: Plenum. doi: 10.1007/978-0-387-85764-0_16.

Schwartz, H. L. (2009). Facebook: The new classroom commons? *Chronicle of Higher Education, 56*(6), B12–13.

Sheldon, P. (2008). The relationship between unwillingness to communicate and students' Facebook use. *Journal of Media Psychology, 20*, 67–75. doi: 10.1027/1864-1105.20.2.6.

Sheldon, P. (2009). I'll poke you. You'll poke me! Self-disclosure, social attraction, predictability and trust as important predictors of Facebook relationships. *Cyberpsychology: Journal of Psychosocial Research on Cyberspace, 3*, article 1.

Sheldon, P. (2013). Examining gender differences in self-disclosure on Facebook versus face-to-face. *The Journal of Social Media in Society, 2*, 89–106.

Sheldon, P. (2014). *Examining student-teacher relationship on Facebook: Theory of reasoned action and uses and gratifications.* Paper presented at the annual meeting of the Association for Education in Journalism and Mass Communication (AEJMC), Montreal, Canada.

Sheldon, P., & Pecchioni, L. (2014). Comparing relationships between self-disclosure, liking and trust in exclusive Facebook and exclusive face-to-face relationships. *American Communication Journal*, *16*(2).

Sorensen, G. (1989). The relationship among teachers' self-disclosive statements, students' perceptions, and affective learning. *Communication Education*, *38*, 259-276. doi: 10.1080/03634528909378762.

Steel, J. L. (1991). Interpersonal correlates of trust and self-disclosure. *Psychological Reports*, *68*, 1319-1320. doi: 10.2466/pr0.1991.68.3c.1319.

Tang, J., & Wang, C. (2012). Self-disclosure among bloggers: Re-examination of social penetration theory. *Cyberpsychology, Behavior, and Social Networking*, *15*, 245-250. doi: 10.1089/cyber.2011.0403.

Thibaut, J., & Kelley, H. (1959). *The social psychology of groups*. New York: Wiley.

Tidwell, L. C., & Walther, J. B. (2002). Computer-mediated effects on disclosure, impressions, and interpersonal evaluations: Getting to know one another a bit at a time. *Human Communication Research*, *28*, 317-348. doi: 10.1111/j.1468-2958.2002.tb00811.x.

Triandis, H. C. (1989). The self and social behavior in different cultural contexts. *Psychological Review*, *3*, 506-520. doi: 10.1037/0033-295X.96.3.506.

Tufekci, Z. (2008). Can you see me now? Audience and disclosure regulation in online social network sites. *Bulletin of Science, Technology and Society*, *28*, 20-36. doi: 10.1177/0270467607311484.

Walker, L. (1984). *The battered woman syndrome*. New York: Springer.

Walther, J., Van Der Heide, B., Kim, S., Westerman, D., & Tong, S. T. (2008). The role of friends' appearance and behavior on evaluations of individuals on Facebook: Are we known by the company we keep? *Human Communication Research*, *34*, 28-49. doi: 10.1111/j.1468-2958.2007.00312.x.

West, R., & Turner, L. H. (2010). *Introducing communication theory:*

*Analysis and application* (4$^{th}$ ed). Boston: McGraw-Hill Higher Education.

Wheeless, L. R., & Grotz, J. (1976). Conceptualization and measurement of reported self-disclosure. *Human Communication Research*, 2, 338–346. doi: 10.1111/j.1468-2958.1976.tb00494.x.

Wheeless, L. R., & Grotz, J. (1977). The measurement of trust and its relationship to self-disclosure. *Human Communication Research*, 3, 250–257. doi: 10.1111/j.1468-2958.1977.tb00523.x.

Worthy, M., Gary, A. L., & Kahn, G. M. (1969). Self-disclosure as an exchange process. *Journal of Personality and Social Psychology*, 13, 59–63. doi: 10.1037/h0027990.

# 2

# 社交媒体与大众传播理论

本章检验了传统的大众传播理论如何发展,从而被运用到我们对社交媒体的理解中。相关理论包括使用与满足理论(uses and gratification)、议程设置理论(agenda-setting)、框架理论(framing)、涵化理论(cultivation)和沉默的螺旋理论(spiral of silence)。

## 使用与满足理论

在大量研究中,使用与满足理论(U&G)(Katz, Blumler, & Gurevitch, 1973)被用于解释人们为什么使用特定的媒介。不同于其他媒介理论(如涵化理论),该理论关注的并不是媒介的内容。它强调受众及他们在选择能够满足需求的媒介时所发挥的积极作用。卡茨、布卢姆勒和古雷维奇(Katz, Blumler, & Gurevitch, 1973)强调我们会选择特定的媒介来满足需求。这些需求可以被分成四类:转移注意力(逃离日常问题)、个人关系(用媒介获得友谊)、个人认同(增强价值感)及追踪(得到能够帮助个体完成某件事的信息)(McQuail, Blumler, & Brown, 1972)。例如,当我们想要笑的时候我们会看戏剧,当我们想要获取信息的时候我们会看

CNN。根据这一理论，人们是自知的，这能够解释人们为什么会去使用媒介。

使用与满足是一种启发式的理论，它激发了大量传统媒体和新媒体（互联网与社交媒体）领域的研究。在过去十年，大量研究检验了 Facebook（Krause, North, & Heritage, 2014; Sheldon, 2008; Smock, Ellison, Lampe, & Wohn, 2011）、Twitter（Chen, 2011; Johnson & Yang, 2009）、YouTube（Hanson & Haridakis, 2008）、Pinterest（Mull & Lee, 2014）、Yelp（Hicks et al., 2012）和博客（Kaye, 2005; 2010）中的使用与满足。因此，一些新的满足被用来解释个体如何使用社交媒体。例如，虚拟社区就是一种用来解释线上沟通的"新"满足（Song, LaRose, Eastin, & Lin, 2004）。社交媒体改变了一些满足的重要性，例如，大多数人通过看电视来获得信息或满足娱乐需求，而通过使用社交网络来保持关系（Sheldon, 2008）。一些社交网站（如 LinkedIn）满足了专业提升的需求，而另一些则允许个人分享表达性的信息（引自 Smock et al., 2011）。克劳斯、诺思和赫里蒂奇（Krause, North, & Heritage, 2014）研究了 Facebook 上对听音乐应用的使用。他们发现了一种新的满足，并且将这一满足命名为爱好消遣型满足。这一动机在使用 Facebook 的其他应用时并不常见。马尔和李（Mull & Lee, 2014）检验了使用者从图片分享型社交网站 Pinterest 中获得的满足。一份验证性因子分析揭示了从图片分享型社交网站中获得使用与满足的五个维度的动机：潮流、创意项目、娱乐、虚拟探索和组织。其中，创意项目和组织动机是新发现。

研究表明，社交媒体使用背后的模式和动机，一部分受人口和性格变量的作用。然而，仅研究社交和个人性格如何影响我们的（普遍意义的）Facebook 使用还不够，因为针对不同类型的人可能要考察不同的特征（Smock, Ellison, Lampe, & Wohn, 2011）。正如斯莫克（Smock）等人指出，"Facebook更多被视为以不同的方式满足不同需求的工具的集合"（p.2323）。在他们的研究中，只有三种动机（放松型娱乐、表达性信息的分享和社会互动）是普遍用途的预测变量，另外六种动机则预测了特定特征下的使用。例如，表达性信息的分享是状态更新和群组使用的预测变量，但不是一对一交流（私聊）的预测变量。因此，那些想要分享个人生活的人不会通过私信分享信息给朋友，他们会公开推送。社会交换动机能显著地预测信息、私聊和圈内推送的使用。专业性的提升能预测圈内推送和私信的使用。

少量研究（Chen, 2011；Johnson & Yang, 2009）探索了Twitter的使用动机。Twitter使用主要包括社交动机和信息动机（Johnson & Yang, 2009）。社交动机包括玩得开心、被娱乐、放松、看他人忙于什么、消磨时间、自由表达自我、和朋友或家庭保持联系、更便捷地沟通以及同时和多人沟通。信息动机包括获取信息、提供或得到建议、学习有趣的事情、认识新的人和分享信息。约翰逊和杨（Johnson & Yang, 2009）发现，使用者最初会出于信息动机来使用Twitter，并且表明是因为Twitter能够方便地定制使用者想要消费的信息流。Twitter用户可以追随特定的用户和新闻机构，从而避免其他新闻网站可能带来的信息超载。

YouTube是另一个被受众中心视角检验的社交媒体平台。汉森和哈里达克斯（Hanson & Haridakis, 2008）发现，观看和分享不同类型的新闻内容会有不同动机。通常，观看较传统类型新闻的人，主要是为了满足信息需求；观看搞笑或讽刺类型新闻的人，主要是为了娱乐。人际传播的动机预测了在YouTube上分享视频新闻的动机。在随后的研究中，哈里达克斯和汉森（2009）检验了个体差异（社交活动、人际交换、控制源、感觉的寻找、创新力和YouTube使用黏度）对在YouTube上观看和分享视频行为的预测。当使用者观看视频是为了获取信息时，他们观看和分享视频则是为了娱乐、共享和社会交换。内在控制源预测了YouTube视频的分享。因此，内在控制型的使用者有更强的自信，并且会使用YouTube一类的网站来改善他们的社交圈和社会生活（Haridakis & Hanson，2009）。人际互动、感觉的寻找、创新力及YouTube使用黏度和观看或分享YouTube视频并不相关。这在某种程度上显露了YouTube作为一个视频分享和社交网站的目的。不同于用户可以在上面与朋友们进行私人性互动（通过私信和聊天）的Facebook，YouTube是一个让用户寻求娱乐和社会交换的地方。一些病毒视频通过其他社交网站在用户中分享证实了这一点。

生产YouTube视频的原因和观看或分享视频的原因不同。YouTube视频的生产者通常喜欢表达自我。莫氏和扬茨（Mosemghvdlishvili & Jansz，2013）对YouTube上关于伊斯兰视频的生产者进行了访谈，研究了他们的不同动机。研究发现了三个主导性动机：和伊斯兰沟通、自我表达、社会认知。和伊斯兰沟通是穆斯林生产者的主要动机，他们谈到，他们试图使用视频博客消除人们对伊斯兰存在的误解。在自我表达方面，所有使用者都表示喜欢YouTube，因为YouTube允许他们更开放地表达自我，并且允许外观和风格上的各种尝试。

少部分研究关注评分型社交网站。希克斯等人（Hicks et al.，2012）研究了使用Yelp的动机。该网站通常被作为餐馆和小型公司的评分系统。Yelp也向小型企业出售广告和赞助商排位。希克斯等人（2012）发现人们使用它的主要原因是分享信息，其次是娱乐与便利。这在一定程度上，与Facebook和YouTube使用动机中的关系维系和娱乐不同。关于另一类打分系统,如 ratemyprofessor.com (RMP)或者 koofers.com 没有引起多少研究兴趣。有研究(Kowai-Bell, Guadango, Little, Preiss, & Hensley, 2011)检验了 RMP 内容在预期和方法上对被评分课程的影响。而研究学生为什么使用这些网站,无论是针对发布者还是阅读者,都会十分有趣。

另一种受使用与满足理论学者大量关注的社交媒体是博客。尽管博客不如社交网站平台受欢迎,但它们仍然存在。博客最初是政治型的,现在更普遍的是日志型的私人博客和线上新闻聚合型的新闻博客(如《赫芬顿邮报》)。最早检验博客使用动机的研究(Kaye, 2005)发现,人们的主要动机包括追踪社会和政治问题,其次是便利,再次是个人满足。凯(Kaye, 2010)在推动博客的使用与满足测量的基础上,将使用动机分为以下几类:便利的信息搜寻、反传统的媒介情感、表达/使用黏性、引导/意见搜寻、博客氛围、个人满足、政治讨论、意见广泛性及特定咨询。近期,阿姆斯特朗和麦克亚当斯(Armstrong & McAdams, 2011)针对年轻用户如何使用博客进行了研究,发现相比以休闲为目的的使用者,以信息搜寻为目的的博客使用者更倾向于相信博客的内容。同样,博客的读者和博客的内容生产者之间也存在差异。纳迪、斯基亚诺、贡布雷希特和斯沃茨(Nardi, Schiano, Gumbrecht, & Swartz, 2004)从博主的汇报中归纳出五种博客写作动机:记录个人生活、提供评论和观点、表达深藏的感情(作为一种发泄方式)、通过写作寻找个人想法、成为社群成员。

## 议程设置理论

根据议程设置理论(McCombs & Shaw, 1972),媒体并不能时时成功地告诉我们怎么想,但它们在一定程度上能成功地告诉我们想什么。该理论指出,人们会认为新闻报道的内容是对自身重要的内容。换言之,"媒介议程"决定"公众议程"。大量研究都关注电视新闻如何塑造公众观念,而大多数基于议程设置视角的研究则关注对反向议程设置和媒介间议程设置的检验。

传统模型解释了媒介议程中的五个因素(记者个人、媒介日程、组织因素、社会机构、文化/意识形态因素;Shoemaker & Reese, 2014),这一模型并不包含社交媒体。众所周知,在大量案例中,公民记者成为新闻的来源,并且经常成为新闻的破坏者。反向议程设置的概念已经成为近几十年的一个研究主题,该概念指的是记者对公共兴趣的回应,以及随之而来的公共议程对媒介议程的推进和影响。金和李(Kim & Lee, 2007)介绍了公共议程设置、媒介议程中反向议程设置的概念。在社交媒体上,公众可以被任何一个有账户的人所代表。古德(Goode, 2009)将之界定为"公民记者"。公民新闻是指普通用户参与新闻活动的实践,比如创作与时事相关的博客、分享照片和视频、发布目击评论(Goode, 2009, p.1288)。公民记者在2010年阿拉伯之春(见第四章)、2009年伊朗革命及2014年弗格森动荡①中均扮演着重要的角色。

罗素·纽曼、古根海姆、江和白(Russell Neuman, Guggenheim, Mo Jang, & Bae, 2014)研究了社交媒体如何促进反向议程设置。尽管罗素·纽曼等人(2014)怀疑社交媒体议程很可能不会设置传统媒体议程,但他们发现社交媒体也没有盲目地跟随传统新闻媒体议程。例如,在生育控制、堕胎、同性婚姻等社会问题上,线上

---

① 弗格森动荡包括2014年8月9日在美国密苏里州弗格森发生的迈克尔·布朗(Michael Brown)致命射击后的抗议。布朗被弗格森的警员射击后死亡。

博客的讨论要多于传统媒体,而传统媒体对经济问题的关注更多。

格力兹温斯卡和博登(Grzywinska & Borden, 2014)的研究关注社交媒体在传统媒体中的议程建设和议程设置,他们选取占领华尔街运动进行个案研究。占领华尔街运动开始于2011年7月13日,该事件的产生受到了阿拉伯之春的启发(见第四章)。抗争者把他们自己界定为"99%的低收入者反对1%的高收入者的贪婪与腐败"(Grzywinska & Borden, 2014, p.2)。就像2010年埃及与突尼斯的抗争者一样,占领华尔街的抗争者通过社交媒体宣布他们下一次的集合地、发布时事通讯或者和其他行动者讨论各类事宜(Grzywinska & Borden, 2014)。在2011年,格力兹温斯卡和博登研究了新闻报纸(《纽约时报》《华盛顿邮报》《洛杉矶时报》)上关于该事件的报道与两个最大的与运动相关的Facebook主页上发布的活动的关系。结果显示,传统媒体在寻找参考来源时,更多选择其他传统媒体,而社交媒体很少这样做;社交媒体主要被其他社交媒体渠道所引用。格力兹温斯卡和博登(2014)得出结论,认为媒体渠道存在一种让用户保持"在褶皱中"(within the fold)的倾向;而社交媒体有时则会为传统媒体进行议程设置,并且影响他们的报道。正如作者所指出的,除了社交媒体,传统记者没有其他关于运动的信源。媒体选择哪些事件、议题或来源着墨的过程被称为议程建设(McCombs, 2004)。

福尔德斯(Volders, 2013)开启了Twitter议程设置研究的先河,探析了Twitter上政治对话中的议程设置理论。研究结果表明,Twitter上与政治相关的活动信息,主要源自包括报纸、杂志、广播和电视在内的传统媒体。而Twitter的例子则揭示了强烈的反向议程设置的效果,即一些Twitter信息的集合如此突出,使得传统媒体将其置于自身的议程中。目前,很少有研究对社交媒体的议程设置和反向议程设置做检验。在社交媒体如何影响议程设置的研究中,发展最成熟的领域是对博客作用的检验。一些研究发现博客促进了传统媒体与公民媒介之间的权力再分配(Drezner & Farrell, 2004; Meraz, 2009)。

更多的研究赞同社交媒体与传统媒体会影响彼此的议程。大众媒介议程对彼此的影响被称为媒介间议程设置(intermedia agenda-setting; Lopez-Escobar, Llamas, McCombs, & Lennon, 1998)。一些研究(Groshek & Groshek Clough, 2013; Ragas & Kiousis, 2010; Sayre, Bode, Shah, Wilcox, & Shah, 2010)表明,社交网站可以成为媒介间议程设置的重要代理方,因为它们能够快速便捷地分享故事、发布新闻。权、李、朴和文(Kwak, Lee, Park, & Moon, 2010)比较了 CNN 头条新闻与 Twitter 上的话题趋势,发现对新闻的破坏最先出现在 Twitter 上。在对 2009 年摩根·哈林顿(Morgan Harrington)失踪事件的相关研究中,阿特威克(Artwick, 2012)发现 Twitter 上的议程是汇总了博客、娱乐及其他种类的非新闻网站信息而设置的。格罗舍克和格罗舍克·克拉夫(Groshek & Groshek Clough, 2013)借时间序列分析,追踪了两个传统媒体领军者(《纽约时报》和 CNN)的媒介间议程设置,以及两个流行的社交网站平台(Facebook 和 Twitter)上最热门的话题。结果表明,社交网站确实具备直接塑造媒体议程的潜力,但仅限于特定话题。而 Twitter 更倾向于追踪传统媒体建构的政治议程,但 Twitter 上的文化报道是社交媒体渠道对传统媒体进行议程设置的一种报道,Facebook 上的文化报道则更清晰地被传统媒体的议程所设置。格罗舍克夫妇(2013)认为,媒体议程研究不仅有主题视角,也存在时间视角。梅斯纳和迪斯塔索(Messner & Distaso, 2008)对《纽约时报》与《华盛顿邮报》六年间的 2 059 篇文章进行了内容分析,发现新闻报纸正大量引用博客作为一种可信的来源(30%—40%的文章引用博客作为来源)。博客也将传统媒体作为重要来源。梅斯纳和迪斯塔索(2008)总结道,传统媒体和博客促成了一个新的来源循环,这使得新闻内容得以在媒介之间来回传递。

# 框架理论

框架理论关注信息内容通过媒介传递并且被受众阐释的方式(Chung & Cho, 2013; Iyengar, 1991)。依据该理论,框架存在着不同的分类。在一部分学者看来(如 Semetko & Valkenburg, 2000),有两种框架:通用的和特定议题的。通用框架涉及的主题更广泛,而特定议题框架则更多地关注细节。通用框架所包含的主题有冲突、经济或人情味故事。特定议题框架则提供了深度的特定议题信息(Wasike, 2013)。还有学者(Iyengar, 1991)区分了情景框架和主题框架。情境框架聚焦于事件和故事,而主题框架强调语境和环境,聚焦于随时间变化的趋势。情景框架关注个体,而主题框架关注某个议题。然而,另一部分学者(Borah, 2014; Cappella & Jamieson, 1997)区分了政治传播中的策略框架和价值框架。策略框架使用战争和竞争的语言(Cappella & Jamieson, 1997),而价值框架会与个体的既有模式产生共鸣,并且强化既有价值。

在政治传播研究中,框架理论已广为人知。该理论具有重要意义,因为媒介对政治议题和内容的选择十分关键;框架会因此影响公众舆论。众所周知,政治舞台中的个体普遍希望建构个人喜好的框架(Bichard, 2006)。近期,政治家已经开始利用社交媒体,直接通过框架和公众沟通。最早关于社交媒体的框架分析始于博客（例如，Bichard，2006；Guillory，2007）。比沙德（Bichard，2006）调查了2004年美国总统选举中候选人网页使用的框架，分析了乔治·布什和约翰·克里网页的四个框架因素（时间、空间、语调和主题）。之后，框架研究开始延伸到Twitter、Facebook和Wikipedia。古德诺（Goodnow，2013）研究了罗姆尼和奥巴马发布在Facebook时间线上的照片，从而理解他们与潜在支持者在交流什么。结果显示，比起奥巴马的团队，罗姆尼的团队会推送两倍数量的图片。这是因为奥巴马已经是总统了，所以他不需要像罗姆尼那样努力树立信服力。另外，罗姆尼的照片主要包含战争英雄和消防员，而奥巴马的图片则以门卫一类的普通人为主。格拉贝和布西（Grabe & Bucy，2009）提出政治家试图塑造选民认知的三种框架：理想的候选人、平民主义的竞选者及确定的败者。

瓦斯科（Wasike，2013）对950条Twitter推文进行内容分析，来检验媒体编辑对所发布文章的框架选择。结果显示，Twitter上的电视编辑相比印刷新闻从业者更倾向于将推文私人化。从主题角度来看，电视社交媒体编辑会倾向推送科技框架的内容，而印刷社交媒体编辑会更多地强调人情味、冲突及经济影响方面的框架。大多数Twitter用户的年轻化趋势和对技术的精通，佐证了科技故事的主导地位（Wasike，2013）。另一个涉及Twitter的框架研究已实施。亨普希尔、库洛塔和赫斯顿（Hemphill, Culotta, & Heston，2013）检验了美国国会的成员如何使用标签，政治家在框架中的参与度，以及哪个议题最受框架影响。他们发现，受框架影响最大的是医疗保健和经济。因为这两个议题是美国共和党与民主党分歧的议题之一，所以这些结果都在意料之中。

钟和曹（Chung & Cho，2013）通过分析美国新闻报纸的相关报道,评价了大众媒体信息和社交网站在中东的角色。研究发现,所有新闻报纸都一致将 Facebook 视为使中东发生变化的主要政治工具。然而,在具体报道的主题上,各报纸间存在着差异。例如,《华盛顿邮报》的报道有着最强的主题性,包括深度故事与专业评论。而《今日美国》则对中东发展报道较少。《国际先驱论坛报》密切关注该议题,主要集中于社交网站在不同威权国家的作用和功能。《纽约时报》则报道了社交网站在威权政权中对民主意识觉醒的作用。

也有学者基于框架理论,对占领华尔街运动进行了研究。研究者(DeLuca, Lawson, & Sun, 2012)认为,Twitter、Facebook 和 YouTube 为行动主义创造了不同于传统媒体的新语境。在该事件中,新闻报纸花费了更长时间去报道占领华尔街的一系列故事,而 Twitter 和 Facebook 从第一天就开始追踪抗争者,并且为其提供支持。德卢卡等人(2012)发现,右翼和左翼的政治博客会对抗争采取不同框架。右翼博客圈普遍未表现出同情,并把抗争建构在"肮脏与危险"的框架中(p.495)。而左翼博客圈则支持运动,强调运动向公众科普美国经济不平等性(DeLuca et al., 2012; Yglesias, 2011)。

针对维基百科的引用在全国性报纸中如何被塑造,梅斯纳和索思(Messner & South, 2011)基于框架理论,对五家美国报纸(《纽约时报》《华盛顿邮报》《华尔街日报》《今日美国》和《基督教科学箴言报》)中维基百科的引用做了内容分析。大多数引用维基百科的文章都发表在《纽约时报》,其次是《华盛顿邮报》。55%文章的框架是中立的,28%是正面的,17%是负面的。梅斯纳和索思(2011)得出结论,通过把维基百科视为一个积极现象和准确的信源,美国报纸提高了线上百科全书的合法性地位。

权和文(Kwon & Moon, 2009)通过分析美国与韩国的报纸和博客上关于美国弗吉尼亚理工大学枪击事件的报道,检验了框架的跨国性差异。尽管他们发现文化差异较小,但结果显示,韩国公众最关心的事实是枪击者是"我们中的一员"。而对于美国人而言,枪击者是"我们中的一员"并不重要。权和文(2009)认为,韩国报纸与公众把这一事件视为对国家声誉的威胁,这可能说明韩国公众倾向于从集体主义视角来理解这一事件。

一些研究表明,通过允许不同声音的出现、边缘群体的出镜,社交媒体提供了塑造事件框架的另一种途径。哈姆迪和戈马(Hamdy & Gomaa, 2012)、哈米斯和马哈茂德(Khamis & Mahmoud, 2013)检验了社交媒体如何塑造埃及示威游行与政治选举的框架。哈姆迪和戈马(2012)发现,在报道2011年埃及示威游行时,不同于报纸把冲突框架作为主导框架,社交媒体采用了人情味框架。哈米斯和马哈茂德(2013)对2012年埃及总统竞选中最热门的五位候选人的Facebook主页进行了内容分析,从而解读他们在竞选前、竞选中和竞选后如何利用Facebook来建构自身的线上形象。讽刺的是,最终成为埃及第一位出自民主竞选的总统候选人,在Facebook上的推文数量最少(Khamis & Mahmoud, 2013)。这表明,除了社交媒体外,其他因素也会影响政治结果(关于其他影响阿拉伯之春因素的更多信息见第四章)。

# 涵 化 理 论

涵化理论(Gerbner & Gross, 1976)表明,大众传媒,尤其是电视,会培养人们对现实的看法。由于电视上存在暴力节目,经常看电视的人会对这个世界产生比现实情况更为暴力的感知。尽管一些学者和媒体专家指出,由于网络对电视形成了冲击,网络可能会加剧乔治·格布纳（George Gerbner）所谓的"残酷世界症候群"（mean world syndrome），但涵化理论则较少引起社交媒体学者的关注。另一个原因是，一些观点相信社交媒体受众并不像格布纳所说的电视受众那么被动。一些研究则发现，传播者在线上发布的信息可能会培养信息接受者的认知。例如，鲍蒂斯塔（Bautista, 2013）认为，传闻存在着循环现象，人们不仅会迅速对自己体验过的餐厅作出负面评价，还会对自己不喜欢的产品作出负面评价。其他用户可能会因此受到影响，不去亲身体验这些餐厅，也不买这些有负面评价的商品。

梅耶尔（Meyer, 2011）认为，涵化理论可以被用来解释女性对个人身体形象的低评价，因为一些女性会把自身和媒体所传达的审美标准进行比较。梅耶尔认为，如今网络正在取代电视成为最普遍的媒介，尤其是在已成年的年轻人群体中。然而，梅耶尔（2011）没有发现比较和低形象评价之间的直接关系。确实，她们并不是通过比较，而是通过选择性分享照片、自尊心和印象管理形成低评价——所以未来的研究应该致力于解释社交媒体在多大程度上可以分享我们对魅力和理想身体形象的认知。

## 沉默的螺旋

根据沉默的螺旋理论（Noelle-Neumann, 1984），如果人们意识到大多数人没有分享他们的看法，他们会更倾向于不谈论这一议题。这些人保持沉默是因为恐惧被孤立。关于沉默的螺旋理论的一个假想是，通过接收来自媒体和个人观察的信息，个体通常会参与评估舆论氛围。如果他们的观点不流行，他们就不会进行分享。诺尔-诺依曼（Noelle-Neumann, 1984）将其称为"准统计官能"

(quasi-statistical sense)。当人们误解了公众舆论,这些观察就被称为"多数无知"(pluralistic ignorance)。

一些研究(例如,Hampton, Rainie, Lu, Dwyer, Shin, & Purcell, 2014; Lee & Kim, 2014; Lemin, 2010)探究了社交媒体是否会改善沉默的螺旋:那些拥有少数观点的群体是否认为能在Facebook或Twitter上更自由地表达他们的观点呢?社交媒体确实可能会吸引那些在其他场合不发声的媒体,但一些研究则持不同假设。汉普顿等人(Hampton et al., 2014)发现,人们在社交媒体上的表现和他们在面对面时的表现相似。他们使用了美国皮尤研究中心具有全国代表性、以1 801个成年人为样本的调查,发现那些使用Facebook和Twitter的人,如果觉察到与朋友的观点一致,会更愿意分享自己的观点。换言之,结果表明,社交媒体并没有给那些总是保持沉默的人提供另类渠道。事实上,当被问及是否愿意讨论爱德华·斯诺登与美国国家安全局的内容时,他们认为相比在Facebook和Twitter上,在家庭聚餐、朋友聚餐或社区聚会上讨论这一话题更轻松。同样,拉明(Lemin, 2010)的结论指出,沉默的螺旋似乎并不会因为社交媒体的环境特性而改变。李和金(Lee & Kim, 2007)的研究也印证了这一点,该研究以来自不同新闻报纸和广播公司的韩国记者为研究对象,来检验他们在Twitter上就争议性话题表达观点的意向。结果显示,那些感知到自己持有少数派观点(更加保守)的记者,相比那些感知到自己持有流行观点的人,更不倾向于表达他们的观点。

然而,有研究(Neill, 2009)发现,当触及硬核(hardcore)问题的时候,社交媒体会对沉默的螺旋理论进行挑战。那些属于少数派的个体会不顾及主流看法而进行发声。在一个对GLBT(男同性恋者、女同性恋者、双性恋者、跨性别者)社群的探索性分析中,尼尔(Neill, 2009)发现,社交媒体给支持者和社群相互之间都提供了入口。未来的研究应该关注另外一些硬核性群体,来了解他们是否利用社交媒体作为一种自我表达与相互支持的另类方式。

## 总　　结

本章讨论了传统大众媒体理论如何被应用于社交媒体研究。在以博客和社交网站为主的社交媒体语境中,被使用最多的理论是使用与满足理论。考虑到使用与满足理论的预设,即媒体受众是积极的,并会因此使用Twitter、Facebook或博客来满足自身需求,该结果并不令人意外。就议程设置理论来看,尽管一些研究表明,该理论在社交媒体的冲击下需要得到更多的修正,但目前还没有充分的证据说明社交媒体带来了广泛的反向议程设置影响,或影响了传统媒体报道(新闻报纸、杂志、广播和电视)。从社交媒体可信度的角度来看,在社交媒体中,任何人都可以成为公民记者,并且在个人账户上发布信息。然而,通过对报道的相互影响,社交媒体促进了媒介间议程设置。作为传统媒体所依赖的信源之一,社交媒体也在议程建设中有重要的作用。此外,另一个在社交媒体领域应得到更多关注的理论是沉默的螺旋理论。"硬核"(少数)群体如何利用社交媒体来表达和组织自身等问题,还有待进行后续研究。

## 参 考 文 献

Armstrong, C. L., & McAdams, M. J.（2011）. Blogging the time away?

Young adults' motivations for blog use. *Atlantic Journal of Communication*, *19*(2), 113-128. doi: 10.1080/15456870.2011.561174.

Artwick, S. G. (2012). *Body found on Twitter: The role of alternative sources in social media agenda setting.* Paper presented at the International Communication Association conference, Washington, DC.

Bautista, V. (2013). *How can cultivation theory be applied in social media.* Retrieved from http://www.socialmediatoday.com/content/how-protect-brands-against-mean-world-syndrome-social-media.

Bichard, S. L. (2006). Building blogs: A multi-dimensional analysis of the distribution of frames on the 2004 presidential candidate web sites. *Journalism & Mass Communication Quarterly*, *83*(2), 329-345. doi: 10.1177/107769900608300207.

Borah, P. (2014). Does it matter where you read the news story? Interaction of incivility and news frames in the political blogosphere. *Communication Research*, *41*, 809-827. doi: 10.1177/0093650212449353.

Cappella, J., & Jamieson, K. (1997). *Spiral of cynicism: The press and the public good.* New York, NY: Oxford University Press.

Chen, G. (2011). Tweet this: A uses and gratifications perspective on how active twitter use gratifies a need to connect with others. *Computers in Human Behavior*, *27*(2), 755-762. doi: 10.1016/j.chb.2010.10.023.

Chung, J. C., & Cho, S. (2013). News coverage analysis of SNSs and the Arab Spring: Using mixed methods. *Global Media Journal: American Edition*, 1-26.

DeLuca, K., Lawson, S., & Sun, Y. (2012). Occupy Wall Street on the public screens of social media: The many framings of the birth of a protest movement. *Communication, Culture, and Critique*, *5*, 483-509. doi: 10.1111/j.1753-9137.2012.01141.x.

Drezner, D., & Farrell, H. (2004). *The power and politics of blogs.* Presented at the American Political Science Association.

Gerbner, G., & Gross, L. (1976). Living with television: The violence profile. *Journal of Communication*, *26*, 172-199. doi: 10.1111/j.1460-2466.1976.tb01397.x.

Goode, L. (2009). Social news, citizen journalism and democracy. *New Media*

& Society, *11*(8), 287–305. doi: 10.1177/1461444809341393.

Goodnow, T. (2013). Facing off: A comparative analysis of Obama and Romney Facebook timeline photographs. *American Behavioral Scientist*, *57*, 1584–1595. doi: 10.1177/0002764213489013.

Grabe, M. E., & Bucy, E. P. (2009). *Image bite politics: News and the visual framing of elections*. Oxford, UK: Oxford University Press.

Groshek, J., & Groshek Clough, M. (2013). Agenda trending: Reciprocity and the predictive capacity of social networking sites in intermedia agenda setting across topics over time. *Media and Communication*, *1*, 15–27. doi: 10.12924/mac2013.01010015.

Grzywinska, I., & Borden, J. (2014). *The impact of social media on traditional media agenda setting theory—the case study of Occupy Wall Street Movement in USA*. Retrieved from http://www.academia.edu/7484515/The_impact_of_social_media_on_traditional_media_agenda_setting_theory._The_case_study_of_Occupy_Wall_Street_Movement_in_USA.

Guillory, B. (2007). *A framing analysis of science and technology weblogs: How is science presented by commentators?* Paper presented at the annual meeting of the International Communication Association, San Francisco.

Hamdy, N., & Gomaa, E. (2012). Framing the Egyptian uprising in Arabic language newspapers and social media. *Journal of Communication*, *62*, 195–211. doi: 10.1111/j.1460-2466.2012.01637.x.

Hampton, K. N., Rainie, L., Lu, W., Dwyer, M., Shin, I., & Purcell, K. (2014). *Social media and the spiral of silence*. Pew Research Center. Retrieved from http://www.pewinternet.org/2014/08/26/social-media-and-the-spiral-of-silence/.

Hanson, G., & Haridakis, P. (2008). YouTube users watching and sharing the news: A uses and gratifications approach. *Journal of Electronic Publishing*, *11*(3), 6. doi: 10.3998/3336451.0011.305.

Haridakis, P., & Hanson, G. (2009). Social interaction and co-viewing with YouTube: Blending mass communication reception and social connection. *Journal of Broadcasting & Electronic Media*, *53*(2), 317–335. doi: 10.1080/08838150902908270.

Hemphill, L., Culotta, A., and Heston, M. (2013). *Framing in social media:*

*How the U. S. Congress uses Twitter hashtags to frame political issues.* Retrieved from http://papers.ssrn.com/sol3/papers.cfm? abstract_id = 2317335" > http://papers.ssrn.com/sol3/papers.cfm? abstract_id = 2317335.

Hicks, A., Comp, S., Horovitz, J., Hovarter, M., Miki, M., & Bevan, J. L. (2012). Why people use Yelp.com: An exploration of uses and gratifications. *Computers in Human Behavior*, *28*, 2274–2279. doi: 10.1016/j.chb.2012.06.034.

Iyengar, S. (1991). *Is anyone responsible? How television frames political issues.* Chicago, IL: University of Chicago Press.

Johnson, P. R., & Yang, S. (2009). Uses and gratifications of Twitter: An examination of user motives and satisfaction of Twitter use. Paper presented at the Communication Technology Division of the annual convention of the Association for Education in Journalism and Mass Communication in Boston, MA.

Katz, E., Blumler, J. G., & Gurevitch, M. (1973). Uses and gratifications research. *The Public Opinion Quarterly*, *37*, 509–623. doi: 10.1086/268109.

Kaye, B. K. (2005). It's a blog, blog, blog, blog world: Users and uses of weblogs. *Atlantic Journal of Communication*, *13*, 73–95. doi: 10.1207/s15456889ajc1302_2.

Kaye, B. K. (2010). Going to the blogs: Exploring the uses and gratifications of blogs. *Atlantic Journal of Communication*, *18*, 194–210. doi: 10.1080/15456870.2010.505904.

Khamis, S., & Mahmoud, A. (2013). Facebooking the Egyptian elections: Framing the 2012 presidential race. *Journal of Arab & Muslim Media Research*, *6*, 133–154.

Kim, S.-T., & Lee, Y.-H. (2007). New functions of Internet mediated agenda-setting: Agenda-rippling and reversed agenda-setting. *Korean Journal of Journalism and Communication Studies*, *50*, 175–205.

Kowai-Bell, N., Guadagno, R., Little, T., Preiss, N., & Hensley, R. (2011). Rate my expectations: How online evaluations of professors impact students' perceived control. *Computers in Human Behavior*, *27*, 1862–

1867. doi: 10.1016/j.chb.2011.04.009.

Krause, A. E., North, A. C., & Heritage, B. (2014). The uses and gratifications of using Facebook music listening applications. *Computers in Human Behavior*, *39*, 71-77. doi: 10.1016/j.chb.2014.07.001.

Kwak, H., Lee, C., Park, H., & Moon, S. (2010). *What is Twitter, a social network or a news media?* Paper presented at WWW 2010, Raleigh, North Carolina.

Kwon, K. H., & Moon, S. (2009). The bad guy is one of us: Framing comparison between the U.S. and Korean newspapers and blogs about the Virginia Tech shooting. *Asian Journal of Communication*, *19*, 270-288. doi: 10.1080/01292980903038998.

Lee, N. Y., & Kim, Y. (2014). The spiral of silence and journalists' outspokenness on Twitter. *Asian Journal of Communication*, *24*, 262-278. doi: 10.1080/01292986.2014.885536.

Lemin, D. (2010). *Public opinion in the social media era: Toward a new understanding of the spiral of silence*. ProQuest, UMI Dissertation Publishing.

Lopez-Escobar, E., Llamas, J. P., McCombs, M., & Lennon, F. R. (1998). Two levels of agenda setting among advertising and news in the 1995 Spanish elections. *Political Communication*, *15*, 225-238. doi: 10.1080/10584609809342367.

McCombs, M. (2004). *Setting the agenda: The mass media and public opinion*. Cambridge, UK: Polity.

McCombs, M., & Shaw, D. L. (1972). The agenda-setting function of the mass media. *Public Opinion Quarterly*, *36*, 176-185. doi: 10.1086/267990.

McQuail, D., Blumler, J. G., & Brown, J. R. (1972). The television audience: Revised perspective. In D. McQuail (Ed.), *Sociology of mass communications* (pp. 135-165). Harmondsworth, UK: Penguin.

Meraz, S. (2009). Is there an elite hold? Traditional media to social media agenda setting influence in blog networks. *Journal of Computer-Mediated Communication*, *14*, 682-707. doi: 10.1111/j.1083-6101.2009.01458.x.

Messner, M., & Distaso, M. (2008). How traditional media and weblogs use each other as sources. *Journalism Studies*, 9, 447–463. doi: 10.1080/14616700801999287.

Messner, M., & South, J. (2011). Legitimizing Wikipedia. *Journalism Practice*, 5(2), 145–160. doi: 10.1080/17512786.2010.506060.

Meyer, M. E. (2011). *Image management on Facebook: Impression management, self-esteem and the cultivation theory.* Presented to the Faculty of the Graduate School of the University of Texas at Austin.

Mosemghvdlishvili, L., & Jansz, J. (2013). Framing and praising Allah on YouTube: Exploring user-created videos about Islam and the motivations for producing them. *New Media and Society*, 15(4), 482–500. doi: 10.1177/1461444812457326.

Mull, I. R., & Lee, S. (2014). "PIN" pointing the motivational dimensions behind Pinterest. *Computers in Human Behavior*, 33, 192–200. doi: 10.1016/j.chb.2014.01.011.

Nardi, B. A., Schiano, D. J., Gumbrecht, M., & Swartz, L. (2004). Why we blog. *Communications of the ACM*, 47, 41–46.

Neill, S. A. (2009). *The alternate channel: How social media is challenging the spiral of silence theory in GLBT communities of color.* Unpublished master's thesis. Retrieved from http://www.american.edu/soc/communication/upload/09-neill.pdf.

Noelle-Neumann, E. (1984). *The spiral of silence: A theory of public opinion—Our social skin.* Chicago: University of Chicago Press.

Ragas, M. W., & Kiousis, S. (2010). Intermedia agenda-setting and political activism: MoveOn.org and the 2008 Presidential election. *Mass Communication and Society*, 13, 560–583. doi: 10.1080/15205436.2010.515372.

Russell Neuman, W. W., Guggenheim, L., Mo Jang, S. S., & Bae, S. (2014). The dynamics of public attention: Agenda-setting theory meets big data. *Journal of Communication*, 64(2), 193–214. doi: 10.1111/jcom.12088.

Sayre, B., Bode, L., Shah, D., Wilcox, D., & Shah, C. (2010). Agenda setting in a digital age: Tracking attention to California Proposition 8 in social media, online news, and conventional news. *Policy & Internet*, 2, 7–

32. doi: 10.1102/1944-2866.1040.

Semetko, H. & Valkenburg, P. (2000). Framing European politics: A content analysis of press and television news. *Journal of Communication*, *50*(2), 93-109. doi: 10.1111/j.1460-2466.2000.tb02843.x.

Sheldon, P. (2008). The relationship between unwillingness to communicate and students' Facebook use. *Journal of Media Psychology*, *20*, 67-75. doi: 10.1027/1864-1105.20.2.6.

Shoemaker, P. J., & Reese, S. D. (2014). *Mediating the message in the 21st century: A media sociology perspective*. New York, NY: Allyn and Bacon.

Smock, A. D., Ellison, N. B., Lampe, C., & Wohn, D. (2011). Facebook as a toolkit: A uses and gratification approach to unbundling feature use. *Computers in Human Behavior*, *27*(6), 2322-2329. doi: 10.1016/j.chb.2011.07.011.

Song, I., LaRose, R., Eastin, M. S., & Lin, C. A. (2004). Internet gratifications and internet addiction: On the uses and abuses of new media. *Cyberpsychology & Behavior*, *7*, 384-393. doi: 10.1089/cpb.2004.7.384.

Volders, S. (2013). *Agenda-setting theory in political discourse on Twitter. Master's thesis*. Retrieved from http://arno.uvt.nl/show.cgi?fid=130756.

Wasike, B. S. (2013). Framing news in 140 characters: How social media editors frame the news and interact with audiences via Twitter. *Global Media Journal—Canadian Edition*, *6*(1), 5-23.

Yglesias, M. (2011, October 13). Wonky protest sign highlights growing inequality. *Think Progress*. [Web log post]. Retrieved from http://thinkprogress.org/yglesias/2011/10/13/343633/wonky-protest-sign-highlights-growing-inequality.

# 3

# 社交媒体心理学

本章将重点介绍社交媒体用户的人格心理学和个体差异。"人格,是理解互联网用户行为的关键要素。"(Amichai-Hamburger,2002,p.1290)随着社交媒体的不断普及,了解哪些人在使用社交媒体显得尤为重要。接下来,本章概述了媒介心理学的相关研究,主要涉及社交媒体使用者和非使用者。

理论上,与在线自我呈现行为相关的人格特质包括:自恋、外向、自我效能感、对归属的需求、对人气的需求。在社交媒体方面,研究者们关注最多的是自恋和外向,因其与社交媒体使用直接相关。其他与社交媒体相关的研究还包括羞怯、孤独和感觉寻求等人格特质。

## 自 我 呈 现

戈夫曼(Goffman,1959)在《日常生活中的自我呈现》一书中,使用了大量夸张的隐喻来描绘社会生活。根据这些隐喻,我们都是表演者,在不同场景中扮演不同的角色。我们有"前台行为"和"后台行为"。当我们遵循标准社会规范时,我们在前台扮演"角色",比如我们在工作中的表现。而我们的后台行为则更随意,比如朋友间的交往(Goffman,1959)。

　　戈夫曼的自我呈现理论已被应用于互联网研究中。沃尔瑟(Walther,1992)用选择性自我呈现(selective self-presentation)来解释当非言语线索缺乏时,言语线索是如何被强化的。线索减少和异步计算机中介传播有助于进行选择性自我呈现(Walther,1996)。这大致上包括了社交媒体上的各种活动,比如更新状态、发布推文、加入小组、点赞,以及在Facebook、Twitter和Instagram上分享个人照片。用户不仅可以自己添加照片,没有照片时也可以允许其他人在照片中标记自己。在隐私设置中,用户则可以限制哪些人能在Facebook中标记他们的照片,这样可以防止出现令人尴尬的照片。总之,通过在线自我呈现,用户能决定在Facebook、Twitter或Instagram上放哪些信息。换句话说,相较于面对面交流,在线上人们可以更容易地掌控自我呈现行为(Ellison, Heino, & Gibbs, 2006)。

　　和其他类型的媒体不同,网络媒体允许参与者共同构建使用环境——这主要通过社会互动来实现(Manago, Graham, Greenfield, & Salimkhan, 2008)。研究还发现,青少年经常尝试不同的网络身份,他们伪装成另一个人,或者只是为了实现在线下生活中受到抑制的某部分天性(Manago et al., 2008; Greenfield, Gross, Subrahmanyam, Suzuki, & Tynes, 2006)。社交网站的吸引力在于,我们可以有无数种方式来展现自我(Manago et al., 2008)。这也是为什么有许多人(例如,Buffardi & Campbell, 2008; Leung, 2013; Mehdizadeh, 2010)认为,自恋者更喜欢较为淡薄的线上关系,因为在这种关系中,他们拥有绝对的自我呈现控制权。社交媒体所提供的"非匿名"(nonymous)(与匿名[anonymous]相对)在线设置,为表达"所希望可能成为的自我"

(hoped-for possible self)提供了一个理想环境。换言之,社交媒体为个人所希望建立的社会化理想身份创造了条件(Mehdizadeh,2010;Zhao,Grasmuck,& Martin,2008)。

## 自恋

自恋是一种人格特质,反映了膨胀的自我意识(Buffardi & Campbell,2008)、对赞美的需求及夸张的自负感(Oltmanns,Emery,& Taylor,2006)。自恋者通常认为,自身优于他人、独一无二、与众不同(Leung,2013)。最初,拉斯坎和霍尔(Raskin & Hall,1979)制作了一个有233个题项的问卷,来衡量自恋程度。随后,拉斯坎和特里(Raskin & Terry,1988)提出了自恋的七个维度:权威、自负、优越感、爱出风头、占有欲、虚荣心和特权感。阿克曼等人(Ackermann et al.,2011)则提出了包含领导力/权威、过强的表现欲、特权感/占有欲三个要素自恋模型。此外,莱昂(Leung,2013)、福斯特和坎贝尔(Foster & Campbell,2007)发现了自恋型人格的四个维度:自视权威或优越、爱出风头、占有欲及渴望虚荣。莱昂(2013)指出,在所有的自恋型人格中,爱出风头的人喜欢尝试各种不同的社交媒体。他们通过社交媒体来显露情感,表达自身的消极情绪和认同感。那些自我感觉优越的人(使用社交媒体)是由认知需求驱动的,而爱慕虚荣的人则希望(通过使用社交媒体)获得认可和重视(Leung,2013)。

研究表明,Facebook正在吸引自恋者。辨识Facebook上的自恋者的最重要指标是他们的头像和社交联系人数(Buffardi & Campbell,2008)。自恋与在Facebook上发布个人照片的频率以及给朋友们的照片评论和点赞的频率有关(Sheldon,2015)。自恋者总会选择能展现自身吸引力的照片作为头像(Kapidzic,

2013）。研究人员（Walther，2007；Winter，Neubaum，Eimler，Gordon，Theil，Herrmann，Meinert，& Kramer，2014）认为，更新Facebook状态是一种最好的印象管理方法，因为个人可以深思熟虑地描写状态。可见，自恋是用户更新社交媒体状态最重要的原因（Winter et al.，2014）。此外，自恋和自我表露的程度成正比，而自我表露被温特（Winter）等人视为一种吸引注意和获得更多"赞"的策略。

门德尔松和帕帕查理斯（Mendelson & Papacharissi，2010）研究了大学生在Facebook相册中的集体自恋。研究者认为，学生们通过Facebook上传照片是一种自觉行为，他们会选择上传特定主题和事件的相关照片，比如高中毕业舞会、体育赛事、万圣节派对、圣帕特里克节和公路旅行。很多照片都记录了仪式或重大事件，比如生日、节日和婚礼。至于照片的主角，大部分都是同性朋友，家人则很少出现在照片中，并且情境和背景都被弱化了。而照片中的表现几乎都是有意的和正式的。许多女性有时会摆出一副挑衅的姿势，有时会摆出和朋友打闹以及互相亲吻的姿势。从美学角度来看，大多数照片都以人物为中心，而且人们都目视镜头。此外，也有很多自拍照。其中，照片的评论功能承担了增强团体凝聚力的作用。

## 外向

另一个备受学界关注的概念是外向。外向被定义为个体的活跃、开朗及参与社会活动的倾向（Winter et al.，2014）。卡尔·荣格最早提出，内向和外向这种人格特征对人的心理机能和交流行为有着重要作用。荣格认为，外向的人一般善于交际、活泼、易相处，并且容易适应社会；而内向的人则完全相反（Acar，2008）。艾森克(Eysenck，1967)指出,在个体的社会化过程中,外向是最重要的人格特质之一。研究发现,外向的人积极、自信、爱寻求刺激、无忧无虑、喜欢占主导、爱冒险(Eysenck, Eysenck, & Barrett, 1985)。麦克罗斯基和里士满(McCroskey & Richmond, 1990)将是否外向视作愿意/不愿意沟通的前提(Acar, 2008)。

针对内向/外向对线上友谊形成的影响,学者们提出了两个不同的假说来解释外向和社交媒体行为之间的关系。富者更富假说(rich-get-richer hypothesis)认为互联网的主要受益者是外向型个体(Kraut, Kiesler, Boneva, et al., 2002)。相反,社会补偿假说(social compensation hypothesis)则认为内向型的人更能从互联网中获益,因为互联网可以弥补内向人群人际交往的不足(McKenna & Bargh, 2000; Valkerburg & Peter, 2007)。就社交媒体而言,研究者发现富者更富假说得到了更多支持(Ong et al., 2011; Sheldon, 2008; Utz, Tanis, & Vermeulen, 2012)。对很多外向的人来说,Facebook是他们社交生活的延伸(Ryan & Xenos, 2011)。Facebook用户(与非用户相比)在外向测试中得分更高(Ryan & Xenos, 2011)。他们将Facebook视作现实人际关系的补充(Seidman, 2013)。此外,Facebook的一些要素尤其吸引外向者,比如状态更新、好友数量及更生动的个人头像(Utz, 2010)。外向的人会加入更多的Facebook群组(Ross et al., 2009)。Facebook对外向的人更具吸引力的一个原因可能是,与以往研究中的聊天室及论坛不同,Facebook是非匿名的,Facebook中的好友在现实生活中通常也互相认识。

然而,关于自恋者与社交网络关系的研究则有不同的发现。比如,麦金尼、凯利和杜兰(McKinney, Kelly, & Duran, 2012)发现,自恋与使用Facebook发布个人信息的频率,没有显著相关关系。而在对大学生的研究中,自恋与使用 Twitter 发布个人信息的频率则呈正相关。作者认为,Twitter 可能是自恋者的理想领域,因为人们可以即刻发送 140 字的消息。不过,他们也承认,社交网站也许仅仅是年轻人的交流方式之一。他们分享个人信息并不是因为自恋,而是因为他们乐于分享这类信息。在最近的研究中,达文波特、伯格曼、伯格曼和费林顿(Davenport, Bergman, Bergman, & Fearrington, 2014)发现,自恋的大学生喜欢用 Twitter 发布状态,而自恋的成年人则喜欢用 Facebook 发布消息。此外,他们还发现,在研究社交媒体和人格特质的关系时,代际差异是很重要的因素。Facebook 伴随着千禧一代长大,然而他们的 Twitter 使用率却更高。相比之下,婴儿潮一代①仍在谨慎地使用 Facebook,试图给他人留下正面的个人形象。

在代际差异方面,特文格、康拉特、福斯特、基思、坎贝尔和布什曼(Twenge, Konrath, Foster, Keith, Campbell, & Bushman, 2008)发现,与 20 年前的同龄人相比,如今的大学生在自恋型人格问卷中获得的分数更高。他们认为这与之前的研究结果相一致。那些研究发现,其他个人主义特质,如自信、行动力、自尊、外向也有所增加。尽管这些特质大多数都是积极的,但有人认为由此所带来的效益只是暂时的。从长远来看,自恋与混乱的爱情关系(Campbell, Foster, & Finkel, 2002)和侵略型行为有关(Bushman & Baumeister, 1998)。

## 羞怯和孤独

学者们所关注的另一概念——羞怯,被定义为"人际交往中的

---

① 指 1946 年至 1964 年出生的美国人。——译者注

不适与抑制,妨碍了个人追求人际关系或职业目标"(Henderson, Zimbardo, & Carducci, 2001, p.1522)。由于缺少非言语线索,一些研究认为,羞怯可能会增加互联网的使用率(Mesch, 2001; Morahan-Martin & Schumacher, 2003)。奥尔等人(Orr et al., 2009)发现,羞怯型用户会在 Facebook 上花费更多的时间。瑞安和塞诺斯(Ryan & Xenos, 2011)则指出羞怯与 Facebook 的使用频率没有显著关系。谢尔顿(Sheldon, 2013a)将 150 名大学生作为研究对象,发现羞怯型学生的 Facebook 好友量少于非羞怯型。羞怯型学生鲜少向 Facebook 好友分享信息,非羞怯型学生则不然。这与以往认为羞怯型用户能在线上互动中获得更多安全感(社会补偿假说)的研究(如 Ward & Tracey, 2004)的假设相抵触。原因可能是,在社会网络环境中,人们很少有机会保持匿名。诸如聊天室、博客和社交网络之类的在线社区,其服务目的各不相同,因而会吸引不同受众。作为社交网络之一的 Facebook,连接的是在现实中有过交集的人们。与聊天室不同,人们可以在聊天室隐藏自己的身份,感受到"被解放了",从而更能放得开。事实上,谢尔顿(2008)发现,那些在面对面交流中感受到焦虑和恐惧的受访者会比其他人更频繁地使用 Facebook,以消磨时间、逃避孤独感。尽管 Facebook 好友很少,但他们仍会频繁地登录。对此,谢尔顿(2008)给出的解释之一是,这些人虽然不会在 Facebook 上自我表露足够多的个人信息以建立新的关系,但他们仍然会经常访问它。

谢尔顿(2008)发现,现实生活中经常发生的事也会发生在 Facebook 上。事实上,学者们已经发现,面对面交流与计算机中介传播(computer-mediated communication, CMC)这两种交流方式既可以共存也可以互相替代(Perse & Courtright, 1993; Rubin & Rubin, 1985)。史密斯和克劳科(Smith & Kollock, 1999)认为,进入虚拟世界时,个人不会抛开现实中的自我,而是会将线下的身份带到线上,并利用线下的身份,来塑造线上的互动和行为。

莫拉汉-马丁和舒马赫(Morahan-Martin & Schumacher, 2003)调查了277名大学生互联网用户,以评估孤独型和非孤独型用户的互联网使用模式差异。研究发现,相比其他人,孤独的人更喜欢线上交流而非线下沟通,他们会更频繁地使用互联网和电子邮件,并且更倾向于通过互联网寻求情感支持。此外,比起非孤独型用户,孤独型用户有更多的网友。但是在Facebook上,情况并非如此。谢尔顿(2013a)针对Facebook上社交孤独和自我表露之间的关系,对美国南部一所大学的150名学生进行了调查。社交孤独被定义为在一个被认可的人际圈中缺乏一席之地或者缺乏一段有意义的友情,通常表现为无聊和排斥感(Weiss, 1973)。谢尔顿(2013a)认为,总的来说,社交孤独的学生要比不孤独的学生在Facebook上表露的信息更少。与那些很容易同他人"合拍"的人相比,社交孤独的人向Facebook好友自我表露的话题更少。这支持了富者更富假说(Kraut et al., 2002)。

在Facebook上,孤独型和羞怯型学生的自我表达似乎有些问题,这样看来最受欢迎的社交网络可能更像一种口碑传播。事实上,这便是社交网络的本质。迪肯-加尔恰(Dicken-Garcia, 1998)称,社交媒体与报纸和电视不同。比起早期的媒体,互联网更强调非正式的人际交往。人们通常用Facebook和朋友们保持联系。使用Facebook最主要的动机是维持关系(Sheldon, 2008),这意味着它满足了人际交流的需要。一般情况下,不同于主要用来同陌生人讨论体育或政治的聊天室和电子公告栏,Facebook并非陌生人聚集和交谈的地方,它是为那些愿意向其他400名实名用户透露个人信息的人创建的。换言之,Facebook的目的是维系与那些我们也会拜访、写信、打电话的朋友们之间的关系。这也许可以解释为什么羞怯型和孤独型学生不愿通过 Facebook 表露自己。

## 追求名声

心理学家指出，通过个性化的新技术进行自我传播，不仅反映了社会中个人主义价值观的转向，也反映出对名声的追求（Greenwood，2013）。乌尔斯和格林菲尔德（Uhls & Greenfield，2012）将渴求名声定义为"在朋友、社群和家庭等直系网络外，寻求积极或消极的公众认同，而不是通过特定的努力获得成就的动机或行为"（p.316）。事实上，根据使用与满足理论，个人的心理需求之一就是对理解和重视的需要（Greenwood，2013）。乌尔斯和格林菲尔德（2012）开展了一个 20 个孩子（9 个女孩，11 个男孩）的焦点小组访谈，受访者年龄在 10—12 岁之间，结果显示，在青春期前的样本群体中，名声排在文化价值的首位。20 个孩子中有 8 位（40%）将名声列为他们未来最想要的东西。这背后的原因是名声与金钱、关注度相关联。他们认为，观察流行电视节目中以成名为导向的内容和发布在线视频来强调这些价值观，二者之间存在潜在的协同效应。在 YouTube 上，人们可以发布关于自身的视频，且拥有广泛受众，所以他们可以得到关于自身名声的真实反馈（Uhls & Greenfield，2012）。乌尔斯和格林菲尔德认为，甚至连网站的标语"展现你自己"（Broadcast Yourself）都在鼓励这种行为。

丹麦男孩本杰明·拉尼耶（Benjamin Lasnier）完全是通过 Instagram 和 YouTube 红起来的，他可以说是"Instafame"（在 Instagram 上爆红）或者说是社交媒体名人最好的例证。拉尼耶最初在 Instagram 上发布自拍照，因为他和贾斯汀·比伯有几分相似，渐渐地开始受人欢迎。截止到 2014 年 11 月，拉尼耶在

Facebook 上已有 540 万粉丝，在 Instagram 上有 120 万粉丝。尽管他的音乐才能备受批评，但索尼公司还是与他签约了。如今，他有了自己的唱片专辑和一系列周边商品。拉尼耶就是乌尔斯和格林菲尔德（2012）研究中观察到的完美例证。通过社交媒体，人人都能成为明星。社交媒体同时也改变了我们对"名声"的看法。

**自我表露**

研究表明，比起线下交流，互联网给那些害羞的人提供了更多在线表露的机会，因为以计算机为中介的渠道可以提供匿名、延迟互动的机会。研究者测试了害羞的人的网络成瘾和在线自我表露行为（Peter, Valkenburg, & Schouten, 2005）。然而，少有研究关注这群人是如何与 Facebook 好友分享信息的。自我表露是友情的一个重要要素，也是测量关系亲密度的指标之一。它被认为是影响关系发展、维护和恶化的重要因素（Levinger & Rands, 1985）。

在面对面交流的情境中，自我表露这一概念一直是学界关注的焦点。随着新技术的发展，该概念也逐渐过渡到以计算机为中介的语境中（如 Joinson, 2001）。自我表露被定义为个体向他人揭示个人信息的过程（如 Berg & Derlega, 1987）。奥尔特曼和泰勒（Altman & Taylor, 1973）提出了社会渗透理论，来描述人们如何在面对面交往中表露信息以发展一段关系。他们提出了自我表露的两个维度：（1）"广度"，即信息表露的总量；（2）"深度"，即自我表露的亲密度。他们认为，通常一段关系在初始阶段是狭窄浅显的。随着关系变亲密，讨论的话题也会变多（广度），并且会对某些话题进行深入讨论（深度）（Altman & Taylor, 1987）。受非言语及情境线索的限制，研究者们（Cho, 2006；Walther, 1992；

1996)认为,在以计算机为中介的环境中,自我表露对线上关系的形成尤为重要。

以往大部分研究认为,害羞的人在网络互动中会享有更多的安全感(例如,Scharlott & Christ,1995;Sheeks & Birchmeier,2007;Stritzke, Nguyen, & Durkin, 2004;Turkle, 1995;Ward & Tracey, 2004)。与之不同的是,谢尔顿(2013a)并未发现证据支持此观点。害羞和孤独的人不会为了弥补面对面交流的缺失,或者为了增加 Facebook 好友,而去使用 Facebook。在谢尔顿(2013a)的研究中,害羞及社交孤独的学生与 Facebook 好友的交流比不太害羞和不孤独的学生更少。同时,那些害羞的学生所拥有的 Facebook 好友及现实好友也比不太害羞的人少。他们可能会因为担心对方不喜欢自己,从而产生焦虑感。所以,害羞的学生不会通过过多的自我表露来发展新关系。但也有研究(如 Baumeister & Leary,1995)发现,害羞的人希望与他人建立关系,且他们对归属感的需求和那些不害羞的人相同。然而,如果他们在线下没有很多朋友,在 Facebook 上也不会有多少。奥尔等人(Orr et al.,2009)的研究结果同样支持了害羞的用户的 Facebook 好友数更少的观点。

**身体意象**

一些实证研究(例如,Lee,2014;Rutledge, Gillmor, & Gillen, 2013)考察了 Facebook 使用和身体意象之间的联系。身体意象是一个多维概念,包括个体如何思考、感受他们的身体以及对身体做出何种举动(Thompson, Heinberg, Altabe, & Tantleff-Dunn, 1999)。以往的研究指出,对人们而言,饮食问题的风险高发期是青春期晚期及刚入大学的过渡期(Dickstein, 1989;

Lorenzen, Grieve, & Thomas, 2004; Striegel-Moore, Silberstein, Frensch, & Rodin, 1989)。饮食失调带来的最常见风险之一就是身体不满意。大约80%的女大学生表示，她们在某一时期都会对自己的身体感到不满(Fitzsimmons-Craft, 2011)。这往往源于同侪间的社会比较(Sheldon, 2013b)。根据费斯廷格(Festinger, 1954)的社会比较理论(social comparison theory)，将自身与他人作比较是人之本能。纵向比较包括与大众媒介中的超模进行比较，而横向比较则是与同辈人进行比较(Sohn, 2010)。该现象在Facebook上尤多。虽然很少有研究关注Facebook这类社交媒体网站如何鼓励社会比较行为，但劳特利奇等人(Rutledge et al., 2013)则认为，作为一种视觉化媒介，Facebook可能会对那些尤为关注自己外表的人有更大的吸引力，因为通过Facebook，他们可以构建出自身希望公众看到的形象。拥有Facebook好友越多的学生，对自己的外表有越积极的评价。那些有更多好友的人也会期待他们发布的照片获得更多的积极评论。劳特利奇等人将之视为人们在Facebook上发布照片的原因之一。结果表明，那些在Facebook上花费时间越少的人则越在乎自身的外表。他们可能担心展现出的形象不够有吸引力。李(Lee, 2014)验证了Facebook上的社会比较倾向(social comparison orientation, SCO)，并研究了大学生的人格特征(如社会比较倾向、自尊、自我不确定和自我意识)对个人在Facebook上进行社会比较的频率的影响。在社会比较倾向上得分越高的学生，越喜欢在Facebook上与他人比较。李将社会比较倾向定义为"个人会关注他人的行为，并根据他人的行为来调整自身的行为的程度"(p.254)。除此之外，李(2014)还发现那些自我确定性较低的学生，通常自尊也较低，但其使用Facebook更频繁，这可能是为了在 Facebook 上与他人作比较。

社会比较研究的局限性在于,大多数人并不愿意承认他们在比较自己和他人(Lee, 2014)。大多数关于社会比较的研究都是横向研究,并不能直接进行因果推断。

**数字鸿沟**

2009年,15%的大学生参与者不是 Facebook 用户。2013年(Ljepava, Orr, Locke, & Ross, 2013),只有8%的大学生没有 Facebook 账户。显而易见,非 Facebook 用户的数量在下降。当下,关注非 Facebook 用户人格特征的研究较少(Ljepava et al., 2013; Ryan & Xenos, 2011; Sheldon, 2012)。瑞安和塞诺斯(2011)指出,通常与 Facebook 用户相比,非 Facebook 用户的自恋和外向程度更低,性格更谨慎,社会孤独感更强。谢尔顿(2012)也发现,非 Facebook 用户往往年纪更大,更易害羞,更具孤独感,社会活动更少且较少寻求刺激类活动。此前有研究(Ryan & Xenos, 2011)表明,非 Facebook 用户群体自恋度较低。相反,列帕娃等人(Ljepava et al., 2013)的研究则发现,非 Facebook 用户在隐性自恋方面(通过超敏感自恋量表[Hypersensitive Narcissism Scale]进行测量)得分更高,这意味着他们只是不太想公开表达自恋。

# 社交媒体与幸福感

学者们还研究了社交媒体技术对个体心理社会幸福感的影响。社交媒体给心理健康可能带来的好处包括社会支持与社会资本(Ellison, Steinfield, & Lampe, 2007)的增加,负面后果则包括社会孤立、面对面交流的减少以及社交媒体成瘾和注意力缺失。

布兰采格(Brandtzæg,2012)将挪威社交媒体用户作为研究对象,指出在媒体推动的"反社交网络"运动中,对社交网站负面影响的表述实际上有误。在2008—2010年三年间,布兰采格(2012)对一个2 000人的挪威在线用户(15—75岁)样本进行了跟踪调查。研究发现,在社会资本的面对面交流、熟人的数量、桥接资本的维度上,社交媒体用户得分比非用户高。换言之,社交网站有助于加强用户与家人、朋友和熟人间的联系。该观点与埃利森等人(Ellison et al., 2007)的研究结果不谋而合。埃利森等人认为,Facebook的使用行为与维护和创造社会资本有关。社会资本即通过人际关系(社会网络)所累积的资源。"桥接型社会资本"包括可能会提供有用信息或新观点,但通常缺乏情感支持的人际"弱关系"或者松散联系。"团结型社会资本"则是通过紧密的、情感密切的关系(家人和亲密的朋友)所建立的。埃利森等人(2007)发现,对很多大学生而言,Facebook是他们同高中好友保持联系的一种途径。事实上,如果使用Facebook更频繁的话,那些低自尊和低满意度的学生会获得桥接型社会资本(穷人变富理论[the poor-get-richer])。社会资本的其他好处还包括情感支持、身心健康的改善、更低的犯罪率(邻里之间互相照顾)、更高的自尊及更高的生活满意度。

此前,布兰采格(2010)界定了五种社交网站用户:散户(sporadics,低级别的社交网站用户)、潜水者(lurkers,使用社交网站但很少发言、少有贡献的人)、社交者(socializers,主要用社交网站与家人朋友进行互动的人)、讨论者(debaters,主要用社交网站进行辩论和讨论的人)以及先进者(advanced,经常使用社交网站且用于各种目的,包括社交、辩论和作贡献的人)。相比于其他类型的用户,社交者有更高的社会资本水平,支持了富者更富假说(Brandtzæg, 2012)。散户和潜水者则比其他用户拥有更少的社会资本。然而,随着时间的推移,研究发现,在社交网站上经常进行讨论和会见新朋友的讨论者和先进者,他们的弱关系均有所增加。但是,社交网站用户会比非用户报告更多的孤独感。布兰采格(2012)认为,这可能是由于男性在社交网站中拥有有价值的关系的比例较低,正如大量研究显示,与男性相比,社交网站对女性社交而言似乎是更为重要的工具(Hargittai, 2007; Sheldon, 2008)。

贝斯特、曼克特洛和泰勒(Best, Manktelow, & Taylor, 2014)还强调,那些拥有高质量关系的人会有更高的幸福感。这也就意味着,在线关系对个体的幸福感有积极影响。社交网站和博客所提供的社会支持则包括增加情感支持、自我表露、减少社交焦虑及获得归属感(Best et al., 2014; Duggan, Heath, Lewis, & Baxter, 2012)。

在线沟通行为与个人幸福感之间的负面关系也已被证实。罗森、奇弗和卡里尔(Rosen, Cheever, & Carrier, 2012)发明了"iDisorder"(自我失序)一词,表示技术使用和心理健康之间的负面关系。其他研究还发现了抑郁症与过度发短信、观看视频片段、玩电子游戏、发邮件及聊天之间的关系(例如,Amichai-Hamburger & Ben-Artzi, 2009; Chen & Tzeng, 2010; 引自 Rosen, Whaling, Rab, Carrier, & Cheever, 2013)。此外,本章前半部分提及的大量研究已揭示了自恋与社交媒体使用之间的关系。自恋受到社交网站的鼓励和推动(Bergman, Fearrington, Davenport, & Bergman, 2011)。由于网络一代(Net Generation)不同的价值观,自恋正在不断增加。除了自恋,过度使用社交媒体还会导致一些人格障碍。比如,罗森等人(2013)对于美国南加州的1 143名成年人(18—65岁)的研究发现,多任务处理会导致抑郁症和狂躁症,以及强迫型和偏执型障碍。

社交媒体还提供了不同的新方式来不断伤害受害者。网络欺凌和网络盯梢只是一部分家长的担忧。网络欺凌被定义为使用互联网、手机或其他技术发送信息或图像,故意伤害和羞辱他人(Levinson, 2013)。大多数网络欺凌的作恶者和受害者均为少女。欺凌者通常喜欢看到自己出现在 YouTube 上,所以他们会在网上发布欺凌视频。近期,一名15岁女孩喝醉后被多名年轻男子伤害。当相关照片被放到网上后,女孩自杀了。网络欺凌与线下欺凌有些不同。首先,线上受害者可以通过注销社交账号来阻止或结束与欺凌者的联系(Menesini & Nocentini, 2009)。其次,在网络环境中,因为无法直接接触受害者,旁观者的潜在同情心会降低。但受限于研究方法,目前只有少数研究充分考察了网络欺凌。因为女性往往更多地被卷入网络欺凌事件中,弗赖斯和古隆(Freis & Gurung, 2013)将 Facebook 中的女性用户作为研究对象,旨在确定促使参与者介入网络欺凌事件的原因(Slonje & Smith, 2008)。结果显示,更具同情心和更外向的人,更有可能介入到网络欺凌的情境中。然而,人们更倾向于用间接干预的方式处理网络欺凌。那些直面欺凌的人,在外向上得分较高,而试图转移话题的人,则在同情心这一项上得分更高(Freis & Gurung, 2013)。另外,上网时间越长越容易增加网络欺凌的风险,传统欺凌和网络欺凌则均与高度的抑郁症状有关(Machmutow, Perren, Sticca, & Alsaker, 2012)。

网络盯梢则是一种持续的、讨厌的在线监视或联系行为,可以看作追踪某个目标到了痴迷的地步(Levinson，2013)。网络欺凌和网络盯梢始于即时通信和聊天室，在这些地方，作恶者可以匿名。之后，他们又移向社交媒体。电子邮件、Facebook、FourSquare 和 Twitter 开始让网络盯梢者可以轻易地跟踪某个人的生活。大多数研究表明，大部分受害者在线上线下都会被同一个人跟踪(Sheridan & Grant，2007)，互联网只是为作恶者控制受害者提供了辅助工具(引自 Welsh & Lavoie，2012)。施皮茨贝格和他的同事们(Spitzberg，Marshall & Cupach，2001)定义了三种类型的网络盯梢：过度亲密(Hyperintimacy)、恐吓(Threat)、影响现实生活(Real-Life Transfer)。过度亲密是指在网络上向受害者不停地传播某些信息，包括夸张的消息和色情图片。恐吓指对网络隐私的侵犯，包括通过电子邮件和 Facebook 消息实施的恐吓。影响现实生活则是指给受害者的生活带来了实质性的伤害(Welsh & Lavoie，2012)。然而，研究发现，网络盯梢者与受害者之间很少有现实接触(Pittaro，2007)。

韦尔什和拉沃伊(Welsh & Lavoie，2012)通过对 321 名女大学生的调研，分析了在社交网站上分享个人信息或线上自我表露与遭受网络盯梢之间的关系。研究者发现，花费在社交媒体上的时间越多，自我表露程度越高，个人遭受网络盯梢的风险会越大。

研究者们通常关注社交网络使用的正负面效应。已有研究实际上证实了社交网站使用所产生的影响既有积极的也有消极的。范多尼克、德海恩斯、德科克和多诺索(Vandoninck，d'Haenens，De Cock，& Donoso，2012)指出，与增加幸福感相关的一个重要变量是，是否以沟通为目的使用网络技术。当以沟通为目的时，社交媒体通常会带来更多的幸福感。

# 参 考 文 献

Acar, A. (2008). Antecedents and consequences of online social networking behavior: The case of Facebook. *Journal of Website Promotion*, *3*, 62-83. doi: 10.1080/15533610802052654.

Ackerman, R. A., Witt, E. A., Donnellan, M. B., Trzesniewski, K. H., Robins, R. W., & Kashy, D. A. (2011). What does the narcissistic personality inventory really measure? *Assessment*, *18*, 67-87. doi: 10.1177/1073191110382845.

Amichai-Hamburger, Y. (2002). Internet and personality. *Computers in Human Behavior*, *18*, 1-10. doi: 10.1016/S0747-5632(01)00034-6.

Amichai-Hamburger, Y., & Ben-Artzi, E. (2009). Depression through technology. *New Scientist*, *204*, 28-29.

Altman, I., & Taylor, D. (1973). *Social penetration: The development of interpersonal relationships*. New York: Holt, Rinehart, & Winstron.

Altman, I., & Taylor, D. (1987). Communication in interpersonal relationships: Social penetration processes. In M. Roloff & G. Miller (Eds.), *Interpersonal processes* (pp. 257-277). London, England: Sage Publications.

Baumeister, R. F. & Leary, M. R. (1995). The need to belong: Desire for interpersonal attachments as a fundamental human motivation. *Psychological Bulletin*, *117*, 497-529.

Berg, J. H., & Derlega, V. J. (1987). Themes in the study of self-disclosure. In V. J. Derlega & J. H. Berg (Eds.), *Self-disclosure: Theory, research, and therapy* (pp. 1-8). New York: Plenum Press.

Bergman, S. M., Fearrington, M. E., Davenport, S. W., & Bergman, J. Z. (2011). Millennials, narcissism, and social networking: What narcissists do on social networking sites and why. *Personality & Individual Differences*, *50*(5), 706-711. doi: 10.1016/j.paid.2010.12.022.

Best, P., Manktelow, R., & Taylor, B. (2014). Online communication, social media and adolescent well-being: A systematic narrative review. *Children &*

Youth Services Review, *41*, 27–36. doi: 10.1016/j.childyouth.2014.03.001.

Brandtzæg, P. B. (2010). Towards a unified media-user typology (MUT): A meta-analysis and review of the research literature on media-user typologies. *Computers in Human Behavior*, *26*, 940–956. doi: 10.1016/j.chb.2010.02.008.

Brandtzæg, P. B. (2012). Social networking sites: Their users and social implications: A longitudinal study. *Journal of Computer-Mediated Communication*, *17*, 467–488. doi: 10.1111/j.1083-6101.2012.01580.x.

Buffardi, L. E., & Campbell, W. K. (2008). Narcissism and social networking web sites. *Personality and Social Psychology Bulletin*, *34*, 1303–1314. doi: 10.1177/0146167208320061.

Bushman, B. J., & Baumeister, R. F. (1998). Threatened egotism, narcissism, self-esteem, and direct and displaced aggression: Does self-love or self-hate lead to violence? *Journal of Personality and Social Psychology*, *75*, 219–229. doi: 10.1016/S0092-6566(02)00502-0.

Campbell, W. K., Foster, C. A., & Finkel, E. J. (2002). Does self-love lead to love for others? A story of narcissistic game playing. *Journal of Personality and Social Psychology*, *83*, 340–354. doi: 10.1037//0022-3514.83.2.340.

Chen, S. Y., & Tzeng, J. Y. (2010). College female and male heavy Internet users' profiles of practices and their academic grades and psychosocial adjustment. *Cyberpsychology, Behavior, and Social Networking*, *13*(3), 257–262.

Cho, S. (2006). *Effects of motivations and gender on adolescents' self-disclosure in online chatting*. Paper presented at the annual meeting of International Communication Association, Dresden, Germany.

Davenport, S. W., Bergman, S. M., Bergman, J. Z., & Fearrington, M. E. (2014). Twitter versus Facebook: Exploring the role of narcissism in the motives and usage of different social media platforms. *Computers in Human Behavior*, *32*, 212–220. doi: 10.1016/j.chb.2013.12.011.

Dicken-Garcia, H. (1998). Internet and continuing historical discourse. *Journalism and Mass Communication Quarterly*, *75*, 19–27. doi: 10.1177/107769909807500105.

Dickstein, L. J. (1989). Current college environments: Do these communities facilitate and foster bulimia in vulnerable students? In L. C. Whitaker and W. N. Davis (Eds.), *The bulimic college student* (pp. 107–133). New York: Hawthorn.

Duggan, J. M., Heath, N. L., Lewis, S. P., & Baxter, A. L. (2012). An examination of the scope and nature of non-suicidal self-injury online activities: Implications for school mental health professionals. *School Mental Health*, 4(1), 56–67. doi: 10.1007/s12310-011-9065-6.

Ellison, N., Heino, R., & Gibbs, J. (2006). Managing impressions online: Self-presentation processes in the online dating environment. *Journal of Computer-Mediated Communication*, 11, 415–441. doi: 10.1111/j.1083-6101.2006.00020.

Ellison, N., Steinfield, C., & Lampe, C. (2007). The benefits of Facebook "friends": Exploring the relationship between college students' use of online social networks and social capital. *Journal of Computer-Mediated Communication*, 12(4). http://jcmc.indiana.edu/vol12/issue4/ellison.html.

Eysenck, H. J. (1967). *The biological basis of personality*. Springfield, IL: Charles C. Thomas.

Eysenck, S. B., Eysenck, H. J., & Barrett, P. (1985). A revised version of the psychoticism scale. *Personality and Individual Differences*, 6, 21–29. doi: 10.1016/0191-8869(85)90026-1.

Festinger, L. (1954). A theory of social comparison processes. *Human Relations*, 7, 117–140. doi: 10.1177/001872675400700202.

Fitzsimmons-Craft, E. E. (2011). Social psychological theories of disordered eating in college women: Review and integration. *Clinical Psychological Review*, 31, 1224–1237.

Foster, J. D., & Campbell, W. K. (2007). Are there such things as "Narcissists" in social psychology? A taxometric analysis of the Narcissistic Personality Inventory. *Personality and Individual Differences*, 43(6), 1321–1332. doi: 10.1016/j.paid.2007.04.003.

Freis, S. D., & Gurung, R. A. R. (2013). A Facebook analysis of helping behavior in online bullying. *Psychology of Popular Media Culture*, 2, 11–

19. doi: 10.1037/a0030239.

Goffman, E. (1959). *The presentation of self in everyday life*. New York: Anchor.

Greenfield, P. M., Gross, E. F., Subrahmanyam, K., Suzuki, L. K., & Tynes, B. (2006). Teens on the Internet: Interpersonal connection, identity, and information. In R. Kraut, M. Brynin, & S. Kiesler (Eds.), *Information technology at home* (pp. 185–200). Oxford: Oxford University Press.

Greenwood, D. N. (2013). Fame, Facebook, and Twitter: How attitudes about fame predict frequency and nature of social media use. *Psychology of Popular Media Culture*, *2*, 222–236. doi: 10.1037/ppm0000013.

Hargittai, E. (2007). Whose space? Differences among users and non-users of social network sites. *Journal of Computer-Mediated Communication*, *13*(1), article 14. Retrieved from http://jcmc.indiana.edu/vol13/issue1/hargittai.html.

Henderson, L. M., Zimbardo, P. G., & Carducci, B. J. (2001). Shyness. In W. E. Craighead & C. B. Nemeroff (Eds.), *The Corsini encyclopaedia of psychology and behavioral science* (pp. 1522–1523). New York: Wiley.

Joinson, A. N. (2001). Self-disclosure in computer-mediated communication: The role of self-awareness and visual anonymity. *European Journal of Social Psychology*, *31*, 177–192. doi: 10.1002/ejsp.36.

Kapidzic, S. (2013). Narcissism as a predictor of motivations behind Facebook profile picture selection. *Cyberpsychology, Behavior, and Social Networking*, *16*, 14–19. doi: 10.1089/cyber.2012.0143.

Kraut, R., Kiesler, S., Boneva, B., et al. (2002). Internet paradox revisited. *Journal of Social Issues*, *58*, 49–74. doi: 10.1111/1540-4560.00248.

Lee, S. Y. (2014). How do people compare themselves with others on social network sites?: The case of Facebook. *Computers in Human Behavior*, *32*, 253–260. doi: 10.1016/j.chb.2013.12.009.

Leung, L. (2013). Generational differences in content generation in social media: The roles of the gratifications sought and of narcissism. *Computers in Human Behavior*, *29*, 997–1006. doi: 10.1016/j.chb.2012.12.028.

Levinger, G., & Rands, M. (1985). Compatibility in marriage and other close relationships. In W. Ickes (Ed.), *Compatible and incompatible relationships* (pp. 309-330). New York: Springer-Verlag.

Levinson, P. (2013). *New new media* (2$^{nd}$ ed.). Penguin Academics.

Ljepava, N., Orr, R. R., Locke, S., & Ross, C. (2013). Personality and social characteristics of Facebook non-users and frequent users. *Computers in Human Behavior*, 29, 1602-1607.

Lorenzen, L. A., Grieve, F. G., & Thomas, A. (2004). Exposure to muscular male models decreases men's body satisfaction. *Sex Roles*, 51(11-12), 743-748. doi: 10.1007/s11199-004-0723-0.

Machmutow, K., Perren, S., Sticca, F., & Alsaker, F. D. (2012). Peer victimization and depressive symptoms: can specific coping strategies buffer the negative impact of cyber victimization? *Emotional & Behavioral Difficulties*, 17(3/4), 403-420. doi: 10.1080/13632752.2012.704310.

Manago, A. M., Graham, M. B., Greenfield, P. M., & Salimkhan, G. (2008). Self-presentation and gender on MySpace. *Journal of Applied Developmental Psychology*, 29(6), 446-458. doi: 10.1016/j.appdev.2008.07.001.

McCroskey, J. C., & Richmond, V. P. (1990). Willingness to communicate: Different cultural perspectives. *Southern Communication Journal*, 56, 72-77. doi: 10.1080/10417949009372817.

McKenna, K. Y. A., & Bargh, J. A. (2000). Plan 9 from cyberspace: The implications of the internet for personality and social psychology. *Personality and Social Psychology Review*, 4, 57-75. doi: 10.1207/S15327957PSPR0401_6.

McKinney, B. C., Kelly, L., & Duran, R. L. (2012). Narcissism or openness?: College students' use of Facebook and Twitter. *Communication Research Reports*, 29(2), 108-118. doi: 10.1080/08824096.2012.666919.

Mehdizadeh, S. (2010). Self-presentation 2.0: Narcissism and self-esteem on Facebook. *Cyberpsychology, Behavior, & Social Networking*, 13, 357-364. doi: 10.1089/cyber.2009.0257.

Mendelson, A., & Papacharissi, Z. (2010). Look at us: Collective narcissism in college student Facebook photo galleries. In Z. Papacharissi (Ed.), *The networked self: Identity, community and culture on social network sites* (pp. 251-272). New York, NY: Routledge.

Menesini, E., & Nocentini, A. (2009). Cyberbullying definition and measurement: some critical considerations. *Journal of Psychology*, *217*, 230-232. doi: 10.1027/0044-3409.217.4.230.

Mesch, G. S. (2001). Social relationships and Internet use among adolescents in Israel. *Social Science Quarterly*, *82*, 329-339. doi: 10.1111/0038-4941.00026.

Morahan-Martin, J., & Schumacher, P. (2003). Loneliness and social uses of the Internet. *Computers in Human Behavior*, *19*, 659-671. doi: 10.1016/S0747-5632(03)00040-2.

Oltmanns, F. T., Emery, E. R., & Taylor, S. (2006). *Abnormal psychology*. Toronto: Pearson Education Canada.

Ong, E. Y. L., Ang, R. P., Ho, J. C. M., Lim, J. C. Y., Gog, D. H., Lee, C. S., et al. (2011). Narcissism, extraversion, and adolescents' self-presentation on Facebook. *Personality and Individual Differences*, *50*, 180-185. doi: 10.1016/j.paid.2010.09.022.

Orr, E. S., Sisic, M., Ross, C., Simmering, M. G., Arsenault, J. M., & Orr, R. R. (2009). The influence of shyness on the use of Facebook in an undergraduate sample. *CyberPsychology & Behavior*, *12*, 337-340. doi: 10.1089/cpb.2008.0214.

Perse, E. M., & Courtright, J. A. (1993). Normative images of communication media: Mass and interpersonal channels in the new media environment. *Human Communication Research*, *19*, 485-503. doi: 10.1111/j.1468-2958.1993.tb00310.x.

Peter, J., Valkenburg, P. M., & Schouten, A. P. (2005). Developing a model of adolescent friendship formation on the Internet. *CyberPsychology & Behavior*, *8*, 423-430. doi: 10.1089/cpb.2005.8.423.

Pittaro, M. L. (2007). Cyber stalking: An analysis of online harassment and intimidation. *International Journal of Online Harassment and Intimidation*, *1*, 180-197.

Raskin, R., & Hall, C. S. (1979). A narcissistic personality inventory. *Psychological Reports*, *45*(2), 590. doi: 10.2466/pr0.1979.45.2.590.

Raskin, R., & Terry, H. (1988). A principal-components analysis of the narcissistic personality inventory and further evidence of its construct validity. *Journal of Personality and Social Psychology*, *54*(5), 890-902.

Rosen, L. D., Cheever, N. A., & Carrier, L. M. (2012). *iDisorder: Understanding our obsession with technology and overcoming its hold on us*. New York, NY: Palgrave Macmillan.

Rosen, L. D., Whaling, K. K., Rab, S. S., Carrier, L. M., & Cheever, N. A. (2013). Is Facebook creating "iDisorders"? The link between clinical symptoms of psychiatric disorders and technology use, attitudes and anxiety. *Computers in Human Behavior*, *29*(3), 1243-1254. doi: 10.1016/j.chb.2012.11.012

Ross, C., Orr, E. S., Sisic, M., Arseneault, J. M., Simmering, M. G., & Orr, R. R. (2009). Personality and motivations associated with Facebook use. *Computers in Human Behavior*, *25*(2), 578-586. doi: 10.1016/j.chb.2008.12.024.

Rubin, A. M., & Rubin, R. B. (1985). Interface of personal and mediated communication: A research agenda. *Critical Studies in Mass Communication*, *2*, 36-53. doi: 10.1080/15295038509360060.

Rutledge, C. M., Gillmor, K. L., and Gillen, M. M. (2013). Does this profile picture make me look fat? Facebook and body image in college students. *Psychology of Popular Media Culture*, *2*, 251 - 258. doi: 10.1037/ppm0000011.

Ryan, T., & Xenos, S. (2011). Who uses Facebook? An investigation into the relationship between the Big Five, shyness, loneliness, and Facebook usage. *Computers in Human Behavior*, *27*, 1658-1664. doi: 10.1016/j.chb.2011.02.004.

Scharlott, B. W., & Christ, W. G. (1995). Overcoming relationship-initiation barriers: The impact of a computer-dating system on sex roles, shyness, and appearance inhibitions. *Computers in Human Behavior*, *11*, 191-204. doi: 10.1016/0747-5632(94)00028-G.

Seidman, G. (2013). Self-presentation and belonging on Facebook: How personality influences social media use and motivations. *Personality and Individual Differences*, *54*, 402-407. doi: 10.1016/j.paid.2012.10.009.

Sheeks, M. S., & Birchmeier, Z. P. (2007). Shyness, sociability, and the use of computer-mediated communication in relationship development. *CyberPsychology & Behavior*, *10*, 64-70. doi: 10.1089/cpb.2006.9991.

Sheldon, P. (2008). The relationship between unwillingness to communicate and students' Facebook use. *Journal of Media Psychology*, *20*, 67-75. doi: 10.1027/1864-1105.20.2.6.

Sheldon, P. (2012). Profiling the non-users: Examination of life-position indicators, sensation seeking, shyness, and loneliness among users and non-users of social network sites. *Computers in Human Behavior*, *28*, 1960-1965. doi: 10.1016/j.chb.2012.05.016.

Sheldon, P. (2013a). Voices that cannot be heard: Can shyness explain how we communicate on Facebook versus face-to-face? *Computers in Human Behavior*, *29*, 1402-1407.doi: 10.1016/j.chb.2013.01.016.

Sheldon, P. (2013b). Testing parental and peer communication orientation influence on young adults' body satisfaction. *Southern Communication Journal*, *78*(3), 215-232. doi: 10.1080/1041794X.2013.776097.

Sheldon, P. (2015). *Self-monitoring and narcissism as predictors of sharing Facebook photographs*. Presented at the Southern States Communication Association conference.

Sheridan, L. P., & Grant, T. (2007). Is cyberstalking different? *Psychology, Crime, & Law*, *13*, 627-640. doi: 10.1080/10683160701340528.

Slonje, R., & Smith, P. K. (2008). Cyberbullying: Another main type of bullying? *Scandinavian Journal of Psychology*, *49*, 147-154. doi: 10.1111/j.1467-9450.2007.00611.x.

Smith, M., & Kollock, P. (1999). *Communities in cyberspace*. London: Routledge.

Sohn, S. H. (2010). Sex differences in social comparison and comparison motives in body image process. *North American Journal of Psychology*, *12*, 481-500.

Spitzberg, B. H., Marshall, L., & Cupach, W. R. (2001). Obsessive relational intrusion, coping, and sexual coercion victimization. *Communication Reports*, *14*, 19-30. doi: 10.1080/08934210109367733.

Striegel-Moore, R., Silberstein, L. R., Frensch, P., & Rodin, J. (1989). A prospective study of disordered eating among college students. *International*

Journal of Eating Disorders, 8, 499–509. doi: 10.1002/1098-108X.

Stritzke, W. G. K., Nguyen, A., & Durkin, K. (2004). Shyness and computer-mediated communication: A self-presentational theory perspective. *Media Psychology*, 6, 1–22. doi: 10.1207/s1532785xmep0601_1.

Thompson, J. K., Heinberg, L. J., Altabe, M., & Tantleff-Dunn, S. (1999). *Exacting beauty: Theory, assessment, and treatment of body image disturbance*. Washington, DC: American Psychological Association. doi: 10.1037/10312-000.

Turkle, S. (1995). *Life on the screen: identity in the age of the Internet*. New York: Simon & Schuster.

Twenge, J. M., Konrath, S., Foster, J. D., Keith Campbell, W. W., & Bushman, B. J. (2008). Egos inflating over time: A cross-temporal meta-analysis of the Narcissistic Personality Inventory. *Journal of Personality*, 76(4), 875–902. doi: 10.1111/j.1467-6494.2008.00507.x.

Uhls, Y. T., and Greenfield, P. M. (2012). The value of fame: Preadolescent perceptions of popular media and their relationship to future aspirations. *Developmental Psychology*, 48(2), 315–326. doi: 10.1037/a0026369.

Utz, S. (2010). Show me your friends and I will tell you what type of a person you are: How one's profile, number of friends, and type of friends influence impression formation on social network sites. *Journal of Computer-Mediated Communication*, 15, 314–335. doi: 10.1111/j.1083-6101.2010.01522.x.

Utz, S., Tanis, M., & Vermeulen, I. (2012). It's all about being popular: The effects of need for popularity on social network site use. *Cyberpsychology, Behavior, and Social Networking*, 15, 37–42. doi: 10.1089/cyber.2010.0651.

Valkenburg, P. M., & Peter, J. (2007). Adolescents online communication and their well-being: Testing the simulation versus the displacement hypothesis. *Journal of Computer-Mediated Communication*, 12(4), article 2.

Vandoninck, S., d'Haenens, L., De Cock, R., & Donoso, V. (2012). Social networking sites and contact risks among Flemish youth. *Childhood*, 19(1), 69–85. doi: 10.1177/0907568211406456.

Walther, J. B. (1992). Interpersonal effects in computer-mediated interaction: a relational perspective. *Communication Research*, *19*, 52 – 90. doi: 10.1177/009365092019001003.

Walther, J. B. (1996). Computer-mediated communication: Impersonal, interpersonal, and hyperpersonal interaction. *Communication Research*, *23*, 3–43. doi: 10.1177/009365096023001001.

Walther, J. B. (2007). Selective self-presentation in computer-mediated communication: Hyperpersonal dimensions of technology language, and cognition. *Computers in Human Behavior*, *23*, 2538–2557. doi: 10.1016/j.chb.2006.05.002.

Ward, C. C., & Tracey, T. J. G. (2004). Relation of shyness with aspects of online relationship involvement. *Journal of Social and Personal Relationships*, *21*, 611–623. doi: 10.1177/0265407504045890.

Weiss, R. S. (Ed.). (1973). *Loneliness: The experience of emotional and social isolation*. Cambridge, MA: MIT Press.

Welsh, A. & Lavoie, J. A. A. (2012). Risky eBusiness: An examination of risk-taking, online disclosiveness, and cyberstalking victimization. *Cyberpsychology: Journal of Psychosocial Research on Cyberspace*, *6*(1), article 4. doi: 10.5817/CP2012-1-4.

Winter, S., Neubaum, G., Eimler, S. C., Gordon, V., Theil, J., Herrmann, J., Meinert, J., & Krämer, N. C. (2014). Another brick in the Facebook wall—How personality traits relate to the content of status updates. *Computers in Human Behavior*, *34*, 194 – 202. doi: 10.1016/j.chb.2014.01.048.

Zhao, S., Grasmuck, S., & Martin, J. (2008). Identity construction on Facebook: Digital empowerment in anchored relationships. *Computers in Human Behavior*, *24*, 1816–36. doi: 10.1016/j.chb.2008.02.012.

第二部分

# 社交媒体的应用

# 4

# 社交媒体与政治

互联网与社交媒体正在改变我们传播、组织和交往的方式。1996年,很多政治竞选人开始创建个人网站,在互联网上发起竞选活动;但直到2004年,公民们才开始利用新的信息技术去了解竞选人。民主党提名的总统候选人之一霍华德·迪恩(Howard Dean),首创了鼓励普通支持者线上参与的方式,初选中为他的竞选筹集到4 000万美元。此外,迪恩也是首位拥有个人博客的候选人。在2008年的美国总统选举中,奥巴马利用社交媒体成功地动员他的支持者,并战胜了迪恩。那些活跃在MySpace、Facebook和Twitter上的年轻选民尤其受影响,仅在Facebook上,他们就为每位政党候选人(贝拉克·奥巴马和约翰·麦凯恩)创建了1 000多个小组。与政治活动相关的社交媒体包括:博客(《赫芬顿邮报》)、微博(Twitter、Google Buzz)、社交网络(Facebook)、职业社交平台(LinkedIn)、视频分享网站(YouTube、Vimeo)和直播网站(Livestream、Justin.TV)。

## 博客与霍华德·迪恩

博客在2004年美国总统选举中扮演了重要角色。通常,政治博客会与其他意识形态相似的博客联系在一起,但在选举期间,它们跟传统媒体讨论相同的问题(Lee,2007)。原因之一在于信源的有限性,因此,高度依赖传统媒体的报道(Lee,2007)。在2004年的选举中,许多职业记者创建了个人博客,一些新闻机构(《纽约时报》、New Republic、ABC)也开始运营博客(Lawson-Borders & Kirk,2005)。博客数量的增长导致"博客"一词被《韦氏词典》(*Merriam-Webster's dictionary*)评选为2004年的年度关键词。而其之所以能吸引大众,是因为它不像传统媒体受控于部分人,可以使人们更自由地表达自我。它们一天24小时开放,全球的人都可以参与到政治讨论中(Mattheson,2004)。马特松(Mattheson)写道:"非职业记者人所运营的新闻博客模糊了新闻和其他表现形式之间的界限。"(p.449)

美国佛蒙特州前州长霍华德·迪恩是首位有效利用博客组织、动员民众的人。在迪恩博客(Dean Weblog)Meetup.com社交网站和数以百计的博客者的帮助下,他进入民主党的总统提名竞选。2002年,斯科特·埃费尔曼(Scott Heiferman)创立了Meetup.com网站,目的是使兴趣相似的人们能够找到彼此,并进行面对面交流(Sifry,2011)。该网站作为e2f(electronic-to-face)平台能让人们在线上交友,并进行线下会面(Weinberg & Williams,2006)。Meetup.com的主要功能就是"为观点相似的陌生人会面安排一个日期、时间与地点的机会"(Gray,2004)。

迪恩曾说,"无法回应"让人们放弃了传统政治,因为传统媒体不存在双向的传播。直到2003年年初,在纽约的见面会上,迪恩才意识到许多支持者都在等待一个与他面对面交流的机会。那些参加过至少一次迪恩见面会的人平均每人向他捐助了154美元(Sifry,2011)。迪恩由此打破了民主党候选人的募款记录。温伯格和威廉斯(Weinberg & Williams，2006)对2004年1—3月中参加过总统候选人见面会的820人进行研究之后发现，他们参加见面会的次数越多，对竞选活动的积极度就越高，从而会更多地参与到捐款、志愿活动、倡导活动中。

迪恩是如何打破募款记录的？草根团体(bottom up)是迪恩和他的竞选经理乔·特里波利(Joe Tripolli)获得成功的秘诀。这种去中心化的团体在大选过程中扮演了重要角色。相反，迪恩的本地、全国志愿者团队只起到极小作用。记者加里·沃尔夫(Gary Wolf，2004)引述了物理学家艾伯特-拉斯洛·巴拉巴西(Albert-Laszlo Barabasi)的观点——"人气孕育更高的人气"(popularity breeds popularity)。换言之，人们会在链接最多的网页上发布更多的链接。这与民主运作的方式有相似之处，"人们互相讨论对他们而言最重要的事情"(David Weinberger，引自Wolf，2004)。沃尔夫认为迪恩的竞选是"愚蠢的网络活动"，它并没有明确的运行目标。但草根团队的网络流程很简易，且很少需要上层指令。例如，迪恩在Meetup.com的支持者们每月都会在全国范围内自发组织见面会。其大部分成员都是中等收入、对政治感兴趣的中年职业人士。为了吸引更年轻的投票人，迪恩的竞选团队还组建了一个名为"Generation Dean"(迪恩一代)的网站。

虽然迪恩在初选环节没有获胜，但他在一年之内从一个政治上默默无闻的人变成了领先者，并在2003年登上了《时代》杂志的封面(Weinberg & Williams，2006)。温伯格和威廉斯(2006)认为，迪恩将互联网应用于政治领域的创举，相当于约翰·肯尼迪在1960年对战理查德·尼克松时，对全国电视联播的创新使用；以及1992年比尔·克林顿对战乔治·布什时，对有线电视的应用。

2011年,罗恩·保罗(Ron Paul)①在Meetup.com拥有约88 000名支持者,有些人甚至认为,茶叶党(Tea Party)诞生于网络。然而,并非每位候选人都能通过Meetup.com获得人气。乔治·华盛顿大学政治、民主和互联网研究院的负责人卡罗尔·达尔(Carol Darr)在2004年的一篇文章中提醒说,网站需要一个有魅力的总统或有魅力的思想吸引人们(Gray, 2004)。

迪恩确实成功地使用互联网获得了早期的支持,但是他最终并未在2004年大选中脱颖而出。四年后,奥巴马用这种方式获得了胜利。2008年,年轻投票人转向Facebook、Twitter等社交网站获取选举新闻。根据民调,年轻投票人的数量达到1972年以来的最高值。在2008年总统竞选中,有46%的美国人使用互联网、电子邮件、手机短信获取相关新闻、分享个人观点、动员他人,这一数字打破了以往的记录(Pew Charitable Trust, 2008)。

## 2008年和2012年美国总统选举中的社交媒体

尽管在2004年美国总统竞选中,博客未得到充分利用,也未能影响到选举结果,但它为公民记者如何组织政治活动提供了方案(Lawson-Borders & Kirk, 2005)。在2008年总统竞选中,包括Facebook和Twitter在内的社交网站则扮演了更关键的角色。

奥巴马的竞选团队雇用了Facebook的联合创始人克里斯·休斯(Chris Hughes),并在他的帮助下创建了myBarackObama.

---

① 罗恩·保罗(Ron Paul),曾于1988年、2008年、2012年三次参加美国总统大选。参见罗恩·保罗个人官方网站 www.ronpaul.com。——译者注

com 网站。此网站提供博客、支持者个人档案、个性化筹款页面、视频、演讲、照片以及事件规划工具。在竞选结束时,该网站已有 200 万活跃用户。奥巴马团队共使用 15 个线上社交网站来推销他们的信息。到 2008 年,奥巴马—拜登的竞选活动在社交网络上共获得了 500 万支持者。相比第二名的约翰·麦凯恩,奥巴马团队对互联网的使用更为充分,除了创建更多 Facebook 群组外,还吸引了更多年轻人(Woolley, Limperos, & Oliver, 2010)。社交网站还能够动员人们通过 YouTube 观看总统竞选的辩论和演讲。最好的例子莫过于奥巴马女孩视频(Obama Girl Video),该视频被《新闻周刊》列为过去十年中十佳"网红"之一。名为"我深情迷恋奥巴马"(I got a crush on Obama)的视频在 2007 年 6 月发布于 YouTube,并引发了病毒式传播。在视频中,一位名为安伯·李·埃廷格(Amber Lee Ettinger)的年轻女性通过唱歌表达她对时任美国参议员奥巴马的爱意。虽然这个视频并未得到奥巴马团队的赞助,但它却获得了数以百万计的观看量,并提升了奥巴马的人气。在整个 2008 年选举期间,无论老少都高度参与到竞选活动中。事实上,那些 Facebook 使用者更可能参与投票,参加集会、聚会,或尝试说服他们的朋友投票,其可能性是非 Facebook 使用者的五倍(Hampton, Goulet, Rainie, & Purcell, 2011)。在竞选活动后,奥巴马从 300 万名线上捐赠者处募捐到超过 5 亿美元,每次捐款的平均金额为 80 美元。其团队共收集了 1 300 万个邮箱地址,发送了超过 7 000 条信息(Levenshus, 2010)。

2012 年,在社交媒体上,奥巴马继续保持着活跃度。他的 Facebook 好友量高达 2 760 万,YouTube 频道订阅者高达 20.7 万,Twitter 关注者超过 1 800 万。其竞选团队所发布的内容数量是米特·罗姆尼的四倍(Pew Research Journalism Project, 2012a)。

Twitter 上的差距尤其明显,平均每天,罗姆尼发布一条推文,奥巴马会发布 29 条。奥巴马的竞选团队建立的社交媒体公共账号也是罗姆尼竞选团队的两倍。然而,两位候选人的 Facebook 和 YouTube 的每条新内容都直接源自竞选活动(Pew Research Journalism Project,2012a),意味着他们均忽略了社交媒体的社交性质,缺乏与普通公众互动。

与 2008 年相比,2012 年奥巴马竞选网站上关注的相关议题数量大幅减少。与之相反的是,罗姆尼则关注了更多不同的议题。两位候选人采取的信息传播策略也不同。根据皮尤研究新闻项目的调查(Pew Research Journalism Project,2012b),罗姆尼团队主要依靠视觉、图形、照片和视频进行信息传播,而奥巴马则更多使用文字。两位候选人将竞选的主要重点议题都放在国内政策上,奥巴马的次要重点是筹资和志愿服务,而罗姆尼则关注竞选活动。虽然在发布的数字信息中,经济是他们竞选头等重要的议题,但经济却并不是选民们最关心的领域。据统计,奥巴马发布的与经济相关的信息平均一条能获得 361 次分享,而关于移民的信息则能获得四倍以上的分享(Pew Research Journalism Project,2012a)。

奥巴马的竞选经理将互联网与草根策略结合。他们向人们发送电子邮件并将信息发布到 MyBO 网上以鼓励支持者组织活动,开展线下会面(Levenshus,2010)。MyBO 网在保持个性化、以用户为中心的同时,还使用了"行动起来"(Take action now)等文字来强调行动的重要性。在 2008 年,奥巴马竞选团队的成员米奇·斯图尔德(Mitch Steward)表示(个人言论,2008 年 12 月 4 日;引自 Levenshus,2010),互联网和网站相当于竞选活动的供给系统,可以吸引志愿者和潜在支持者,所有注册的人都会以个人形式参与其中。

以媒介史观之,奥巴马对社交媒体的使用并非一项创举。富兰克林·罗斯福曾借广播进行炉边谈话(Fireside Chats),肯尼迪则使用电视(来获取民众的支持)。但社交媒体或许是贝拉克·奥巴马成为美国总统的关键原因。保罗·莱文森(Paul Levinson)在《新新媒介》(*New New Media*)一书中提到,社交媒体帮助奥巴马获得了选举的胜利。

## 社交媒体是否影响了政治竞选?

在竞选活动中使用社交媒体的好处如下:
- 获取相关政治资讯;
- 从单向的信息流动方式转向同侪对同侪(peer-to-peer, P2P)的分享方式(Jenkins, 2006);
- 社交媒体"与朋友分享"(tell-a-friend)的特点使人们能够将竞选信息转发给朋友;
- 成本收益率变高:一条 YouTube 视频是免费的,一则电视广告耗费百万;
- 降低了时间与精力上的投资,快速获得回报;
- 人们与自己的朋友、家人分享文章,提升了与候选人的亲密度;而在电视上,人们对候选人有距离感;
- 与电子邮件相比,相对不打扰公众。

宝川和卡蒂(Takaragawa & Carty, 2012)讨论了"与朋友分享"的现象如何改变个人信息获取方式。例如,比起电视广告,人们对好友通过社交媒体分享的视频更感兴趣。这是因为人们更愿意相信他们认识并私下有联系的人。在另一项研究中研究者(Noort, Antheunis, & van Reijmersdal, 2012)发现,当社交网络使用者从联系密切的对象（家人、朋友）那里收到市场营销活动的信息时，他们会采取更积极的态度，更愿意将之转发给社交好友。社交媒体的另一优势体现在定向政治活动中（Christopoulos，2013）。由于竞选团队有可能获得社交媒体使用者的社会人口信息，因此观点也可以像产品一样进行售卖。因此，选民的分类可以极其精细（Christopoulos，2013）。例如，2012年，奥巴马竞选团队为26个不同的选民类别量身定制了竞选信息。其他学者（Bekafigo，Cohen, Gainous, & Wagner, 2013）指出，尽管社交媒体并非旨在成为政治门户网站，却已经适应了这一目的。奥巴马总统和国会议员正用Twitter向他们的支持者发布简短的推文。互联网和社交媒体已经成为全国竞选活动，包括传播、造势、募捐和组织的重要部分（Gainous & Wagner，2011）。此外，社交媒体还让非传统的候选人有机会竞争公共岗位，否则他们将处于劣势（Allison，2002；Wagner & Gainous，2009）。均等化理论（equalization theory，Barber，2001）认为互联网是一个良性的民主化实体，有助于消除选举过程中对某些团体和党派有利的壁垒。

拉马尔和铃木-兰布雷希特（LaMarre & Suzuki-Lambrecht，2013）调研了2010年美国国会选举中Twitter作为公共关系传播工具的效能。研究结果表明，国会选举中，Twitter使用者的获胜概率比非使用者更高。他们还发现，候选人的推文量并不会帮助他们获得选举，而关注量显著增加了他们的获胜概率。拉马尔和铃木-兰布雷希特总结道，在选举中，只有在拥有较多积极参与的关注者时，Twitter才能产生效果。相似的，莱文舍斯（Levenshus，2010）检验了2008年总统竞选中奥巴马—拜登团队对社交媒体的利用。他们发现草根行动策略（grassroots activism strategies）通过社交媒体得以成功实施，提升了选民对候选人的投入。伊芙柯

(Ifukor，2010)对2007年尼日利亚选举期间的245条博文和923条推文展开了研究。结果显示，Twitter是参与型政治(paticipatory politics)的另一种途径。容赫尔(Jungherr，2014)分析了2009年德国联邦议院选举期间评论政治的推文，发现推文的内容混合了政治新闻的多种逻辑。有时采用了传统新闻媒体报道政治人物的逻辑，有时采用了社交媒体的逻辑。容赫尔(2014)引用了安德鲁·查德威克(Andrew Chadwick)对这种现象的命名——"混合媒介系统"(hybrid media system)(Chadwick，2013)。查德威克认为，传统媒体和社交媒体上的政治传播互相关联、互相依存。帕帕查理斯和德·法蒂玛·奥利韦拉(Papacharissi & de Fatima Oliveira，2012)则指出，Twitter报道事件的方式与传统媒体不同，Twitter上的新闻主要以主体经验、观点和情感为基调。此外，容赫尔(2014)还发现推文具有帮助非传统政党动员、进行大量讽刺性评论的新作用。在其他方面，Twitter的内容仍遵循传统媒体的逻辑：个性化、竞争、赛马报道和资讯索引(Jungherr，2014)。

## 使用社交媒体的缺点

电视曾是不同政治光谱(political spectrum)的人们都观看的社区性媒介；而在数字时代，互联网为媒介消费者提供了选择自己喜欢的新闻的机会。结果是，人们与意识形态相同的人互动(Warner & Neville-Shepard，2011)。沃纳和内维尔-谢泼德(Warner & Neville-Shepard)认为，数字媒体导致了媒介受众的分裂和分化。根据他们的描述，"由于许多被动的、温和的声音从对话中消失，这种'选择参与/退出'的政治模式('opt in/opt out' version of politics)变得更为分化"(p.202)。他们分析了迪恩的

blogforamerica.com 上的博文，发现它们显示了敌对的主题，比如"夺回我们的国家"（take back our country）。他们还发现博文常常提到美国独立战争，并注意到迪恩的竞选团队攻击传统媒体，指控它们"使政治成为精英主导的事件"（p.208）。其他研究者（如Sunstein，2007）也引用迪恩的博客作为敌对的例证，认为这种分裂会引起分化，甚至带来极端主义、憎恨和暴力。与迪恩的博客相反，奥巴马网站的博客没有敌对的话语。对奥巴马的支持者而言，麦凯恩"仅仅是不太理想的候选人"（p.211），而非敌人。

# Facebook 政治群组

政治总是涉及网络化。政治家因此利用 Facebook、Twitter 和 YouTube 来接触分散的社群，并为了某一目的，在特定的区域、特定的时间，向政党成员窄播（narrowcast）信息（Elmer & Langlois，2013）。但是，Facebook 政治群组存在一定的局限性。首先是错误共识效应（false consensus effect）（Woolley et al.，2010）——人群中的大部分人分享同样的观点（Marks & Miller，1987）。其次，有研究表明，总体而言，Facebook 政治群组上的信息内容和讨论的质量偏低（Conroy，Feezell，& Guerrero，2012）。并且，在这些群组中表达的观点可能与其他媒介上的相似（Woolley et al.，2010）。李（Lee，2007）对 2004 年美国总统竞选期间的博文和主流媒体新闻进行内容分析，发现博文与主流媒体的议程相差无几。

社交媒体和新信息技术被许多政府当作维护公共关系的工具（Hong，2013），成为"电子政府的核心部分"（Bertot，Jaeger，& Hansen，2012）。同时，新信息技术也改变了政府与公民之间的互动方式。以往的研究表明，高质量的政府网站能提高政府透明度，为网站和政府累积可信度;能让人们更容易知悉政府政策和活动信息(Searson & Johnson, 2010;引自 Hong, 2013)。它能使公众更加信任政府,而许多人认为公共信任(public trust)正是民主的重要基石(Sadeghi, 2012)。根据 2011 年电子政务调查(Norris & Reddick, 2011),大部分美国地方政府使用了至少一种社交媒体,其中使用最普遍的媒介包括 Facebook、Twitter 和 YouTube。洪(Hong, 2013)分析了超过 2 000 位美国公民的数据,研究使用政府网站或与政府的社交媒体账号联系的个人经验,是否会影响人们对政府—公众关系的认知。洪发现,在地方和州的层面,人们借助社交媒体与政府联系的经验与对政府的信任度呈正相关。那些利用社交媒体与当地政府互动的人对政府的信任度更高,但这一结论不适用于联邦政府网站。洪(Hong, 2013)和其他学者(Schario & Konisky, 2008)均认为,公民对当地政府更为信任的原因在于,他们拥有与当地政府直接联系的经验,而全国新闻总是聚焦于联邦政府令人不愉快的作为上。最后,研究(Norris & Reddick, 2013;Waters & Williams, 2011)指出,尽管研究者希望社交媒体能实现政府与公众的双向传播,但目前,政府对新技术的使用仍局限于单向传播。

对于政府—公民关系,林德斯(Linders, 2012)认为,相比于线下形式,社交媒体提供了诸多益处,如更易于吸引兴趣相似的成员、交换信息、不需要通过层级去经营、管理群组。林德斯(2012)、约翰斯顿和汉森(Johnston & Hansen, 2011)认为公民合作生产越来越与技术的先进性相关。在当下,合作生产有三种类型(Linders, 2012):"公民至政府"(citizen to government)、"政府至公民"(goverment to citizen)和"公民至公民"(citizen to citizen)。"公民至政府"的范畴包括政府通过 Facebook、Twitter

等社交媒体向公众征询意见,如奥巴马总统的 change.gov 倡议活动。这一类别包括了服务监督和公民报道(Linders, 2012),公民能够有效便利地与政府分享知识。例如,他们可以拍下逃犯、自然灾害造成损坏的照片,帮助官员采取行动。第二种类型是"政府至公民"或"作为平台观点的政府"(government as a platform idea)。在这种类型中,政府会向公民提供健康风险信息、可供申请的政府项目、邻近地区的犯罪情况等高度个人化的信息,帮助公民作出决策。第三种类型是"公民至公民"或"DIY 政府"(do-it-yourself government),社交媒体帮助公民更有效地自主安排、自我组织,直接帮助彼此,并在政府未良好运作时协调彼此的行动(Linders, 2012)。阿拉伯之春即为例证。合作生产的支持者认为这些行动有助于累积社会资本,推动公民社会发展(Torres, 2007),促进创新并帮助经济困难群体和弱势群体参与到行动中(Bovaird, 2007)。

## 社交媒体与阿拉伯之春

许多学者认为社交媒体是阿拉伯世界一些国家民众推翻威权政权的"催化剂"(Frangonikolopoulos & Chapsos, 2012; Zhuo, Wellman, & Yu, 2011)。2010 年,阿拉伯之春以突尼斯为起点,随后蔓延到其他中东和北非国家。传统媒体在这些威权政权内受到高度管控和限制,社交媒体却让人们能够自由地、不受限制地分享对政府的不满。当伊朗总统马哈茂德·艾哈迈迪-内贾德赢得 2009 年总统选举时,参与"绿色革命"(Green Revolution)的伊朗民众利用社交媒体记录了示威实况(Ali & Fahmy, 2013)。由于 Twitter 在这些抗议活动中被广泛使用,不少学者认为"公民记者"

就诞生于2009年伊朗总统选举后（Human Rights Watch, Iran, 2010；引自 Ali & Fahmy, 2013）。其中最为知名的是哲学系学生内达·阿迦-索尔坦（Neda Agha-Soltan）在旁观德黑兰 Kargar 街头抗议时胸部中弹，她的死亡视频被传到 YouTube 上。

得知伊朗的示威活动后，苦于威权政权和贫困的埃及和利比亚的社会活动人士开始号召民众反对长达30年的穆巴拉克政权（Ali & Fahmy, 2013）。2011年1月，埃及掀起"Facebook 革命"（Facebook Revolution）。卓等人（Zhuo et al., 2011）认为是社交媒体使埃及社会发生转型。在2010年，虽然只有29%的埃及人能够上网，但那些受过教育的"Facebook 一代"决定反对传统主义、专制及社会和经济动荡。埃及革命也常被称为"三重革命"（Triple Revolution）（Zhuo et al., 2011），它包括：

（1）社交网络的兴起；

（2）实时网络（instantaneous Internet）的繁荣；

（3）随时可用的手机。

在阿拉伯之春发生前，阿拉伯世界的年轻人已经开始使用社交网络。手机价格低廉，便于隐藏，同时能用于拍照记录、与他人分享抗议示威的现场。在突尼斯进步青年运动（Progressive Youth of Tunisia）中，虽然社会活动人士通过互联网传播信息，创建相关网页链接，但他们仍不得不依靠其他渠道向无法上网的民众传达信息。很多人还是通过面对面交流或口头传播了解抗议情况。还有人通过短信向公众分享现场图片。卓等人（2011）用"网络个人主义"（networked individualism）这一术语来形容群体控制减弱、自主性增强的网络化社会。

在阿拉伯之春中，社交媒体的另一个重要作用是帮助人们建立归属感，降低孤立感。Facebook 页面和 YouTube 视频"我们都是哈立德·赛义德"(We are all Khaled Said)的创建最能说明这一点。某晚,埃及年轻小伙哈立德·赛义德在网吧被警察拷打至死,原因是他发布了一组记录警察腐败的新闻片段(Ali & Fahmy, 2013)。赛义德的死亡引起了民主人士的愤慨。中东、北非的谷歌执行官瓦伊尔·高尼姆(Wael Ghonim)为此创建了相关 Facebook 页面,吸引了数以千计的粉丝,从而引起国际组织关注并向埃及政府施压,要求两名涉案警察接受审讯。这一页面成为埃及众多抗议活动的催化剂。有人为赛义德制作了一段 YouTube 视频,用一系列照片对比了赛义德的幸福生活和悲惨死亡(Preston, 2011)。尽管赛义德与其他被杀的抗议者无异,但通过社交媒体,他的故事广为流传,"我们都是哈立德·赛义德"Facebook 页面上哈立德·赛义德的照片成为这位烈士的视觉表征(Halverson, Ruston, & Trethewey, 2013)。

根据霍尔(Hall, 2012)的研究,在革命发生后的六个月内,埃及的 Facebook 用户从 45 万增加到 300 万。其中仅有 15% 的用户将之用于娱乐,人们主要通过 Facebook 增强对抗议活动的认识(31%),组织活跃人士参加抗议活动(30%),并向全世界传播信息(24%)(Salem & Mourtada, 2011)。2011 年 2 月,利比亚民众受到伊朗、埃及抗议活动的鼓舞,开始抗议卡扎菲总统的统治。在抗议活动爆发时,卡扎菲迅速切断了包括电视、互联网在内的所有通信方式,但利比亚民众仍然用照相机记录下抗议活动。Twitter 成为推动卡扎菲政权终结的关键工具。卡扎菲于 2011 年 10 月被枪杀。

## 社交媒体催生了阿拉伯之春?

历经长期的争议,学者们基本达成共识,认为社交媒体在阿拉伯之春中起了软决定性的作用。赫斯特（Hirst，2012）认为，新闻是历史的"初稿"，我们不能武断地将抗议活动与社交媒体决定论联系在一起。他提醒说，在突尼斯、巴林、埃及、也门、利比亚和叙利亚发生的抗议事件，包括民主运动和工人运动，早在2011年1月前已开始。"这是守旧的工人阶级支持民主化运动繁荣发展。"（Lee & Weinthall，2012，p.283）赫斯特（2012）认为，由于西方记者对阿拉伯世界的情况知之甚少，未曾设想抗议行动的爆发。因此，当抗议发生后，他们主要依赖互联网上的公民新闻了解情况。这种"便利偏误"（bias of convenience，普林斯顿历史学家爱德华·田纳[Edward Tenner]发明了这一术语）造成了记者间的群体思维和从众心理（Tiffen，1989）。赫斯特（2012）指出，便利偏误和软决定论仍然存在于当今的新闻业中，并导致人们相信社交媒体催生了阿拉伯之春。换言之，公民记者成为全球媒体的通讯员。

麦卡弗蒂（McCafferty，2011）讨论了阿拉伯之春运动中的行动主义与懒人行动主义（slacktivism）。懒人行动主义是指人们愿意为抗争行动的初衷点赞，却不会真正采取实际行动。例如，在一次线上请愿书活动中，支持者们会直接将请愿书的模板复制粘贴，但从不提出他们自己的想法。麦卡弗蒂认为，这种行为并非行动主义，并由此强调了传统社会运动和现代社会运动的差异性。传统社会运动主要由有着强关系的人们组织（教会成员、班级同学），而经由社交媒体发生的社会运动则有赖于弱关系。麦卡弗蒂总结，在阿拉伯之春中，人们仅仅是用社交媒体传播抗议现场实况，而行动主义无论是过去或将来都与行动者本身有关。格拉德韦尔（Gladwell，2010）也提出弱关系无法生成高风险的抗争行动。他以美国民权运动为例，指出社会运动的发生有赖于强关系或强个人纽带。

阿里和法赫米（Ali & Fahmy，2013）持相似观点，认为社交媒体并非伊朗、利比亚、埃及发生革命的唯一因素。在运筹上，行动主义者使用社交媒体更方便，但口头传播仍然是主要的媒介。他们认为作为民众信息的主要来源，传统媒体在阿拉伯之春中仍然占据主流。例如，在Dream TV采访了赛义德Facebook页面的创始人并在半岛电视台播出之后，这一事件才开始受到大量关注。尽管埃及革命又被称为"Facebook革命"，但阿里和法赫米认为，公民只是将社交网络视作提醒国际媒体的另一种工具罢了。在伊朗、埃及和利比亚，口头传播是信息散布最有力的方式（Ali & Fahmy，2013）。而社交媒体是革命的助推器（Halverson et. al，2013）。

传播学者、技术决定论①者马歇尔·麦克卢汉在他的早期作品（《理解媒介》，1964年）中提出，技术即传播，媒介即讯息，形式决定了内容。麦克卢汉将传播技术视作社会变革的引擎（Carey，1967）。在硬决定论中，A是导致B的必要条件，例如头上淋水是头变湿的必要条件；而在软决定论中，可能有其他促使事件发生的条件（Levinson，2013）。社交媒体并非阿拉伯之春事件的决定性因素，但是它推动了抗议行动的发生。"我们都是哈立德·赛义德"Facebook页面的创建者瓦伊尔·高尼姆将埃及2.0革命比作一个维基百科页面，每个人对之都有所贡献，但没有任何人担任领导的角色，人人都是公民记者。阿里和法赫米（2013）立足公民记者的把关行为展开研究。把关是新闻选择的过程，把关人将符合一定标准的内容转换为新闻，并纳入传播管道中（White，1950）。

---

① "技术决定论"最早由美国社会学家凡伯伦（Thorstein Veblen，1857—1929）提出。他假设技术是社会、文化发展的动力。

阿里和法赫米认为,用户生成的内容并未威胁到传统媒体的地位,反而受制于专业记者所倡导的把关原则。同时,传统媒体可以从社交媒体上挑选出具有新闻价值的内容,这些内容往往是一些富有人情味的故事。

社交媒体连接了整个阿拉伯世界,让人们共同为民主和变革努力(Frangonikolopoulos & Chapsos,2012)。普通公民可以通过社交媒体表达对政府的不满、传播消息并提高对公民社会的意识,从而传播勇气,互相支持。贝克特(Beckett,2011)认为,社交媒体让人们产生了坚定感,使他们走得更远。它将人们联合起来,赋予人们权利,为共同的目标努力,并最终引起了世界的关注。

林赛(Lindsey,2013)认为,"在叙利亚,社交媒体为人们获得国际同情和支持提供了渠道"。人们通过手机、Facebook 和 YouTube 分享的非专业视频,让全世界看到了巴沙尔·阿萨德政权对它的公民所行之事。但是,林赛承认中东地区的社交媒体普及率较低,70%的人仍通过电视节目获取信息。

# 参 考 文 献

Ali, S. R., & Fahmy, S. (2013). Gatekeeping and citizen journalism: The use of social media during the recent uprisings in Iran, Egypt, and Libya. *Media, War, & Conflict*, 6(1), 55–69. doi: 10.1177/1750635212469906.

Allison, J. E. (2002). *Technology, development, and democracy: International conflict and cooperation in the information age, SUNY series in global politics*. Albany: State University of New York Press.

Barber, B. R. (2001). The uncertainty of digital politics. *Harvard International Review*, 23(1), 42–47.

Beckett, T. (2011). *After Tunisia and Egypt: Towards a new typology of media and networked political change*. Retrieved from http://blogs.lse.ac.uk/

polis / 2011 / 02 / 11 / after-tunisia-and-egypt-towards-a-new-typology-of-media- and-networked-political-change /.

Bekafigo, M., Cohen, D., Gainous, J., & Wagner, K. (2013). State parties 2.0: Facebook, campaigns, and elections. *The International Journal of Technology, Knowledge, and Society, 9*, 99-112.

Bertot, J. C., Jaeger, P. T., & Hansen, D. (2012). The impact of policies on government social media usage: Issues, challenges, and recommendations. *Government Information Quarterly, 29*, 30-40.

Bovaird, T. (2007). Beyond engagement and participation: User and community coproduction of public services. *Public Administration Review, 67*(5), 846-860.

Carey, J. W. (1967). Harold Adams Innis and Marshall McLuhan. *The Antioch Review, 27*(1), 5-39. doi: 10.2307/4610816.

Chadwick, A. (2013). *The hybrid media system: Politics and power.* Oxford, England: Oxford University Press.

Christopoulos, D. (2013, July). *Does social media impact political campaigns? Volatility and salience in political discourse.* Presented at the NCSL Symposium for Legislative Leaders, Scottish Parliament.

Conroy, M., Feezell, J., & Guerrero, M. (2012). Facebook and political engagement: A study of online political group membership and offline political engagement. *Computers in Human Behavior, 28*(5), 1535-1546. doi: 10.1016/j.chb.2012.03.012.

Elmer, G., & Langlois, G. (2013). Networked campaigns: Traffic tags and cross platform analysis on the web. *Information Polity, 18*(1), 43-56. doi: 10.3233/IP-2011-244.

Frangonikolopoulos, C., & Chapsos, I. (2012). Explaining the role and impact of the social media in the Arab Spring. GMJ: *Mediterranean Edition, 8*(1), 10-20.

Gainous, J., & Wagner, K. (2011). *Rebooting American politics: The Internet revolution.* Lanham, MD: Rowman and Littlefield.

Gladwell, M. (2010, October 4). Small change: Why the revolution will not be tweeted. *The New Yorker.* Retrieved from http://www.newyorker.com / reporting/2010/10/04/101004fa_fact_gladwell? currentPage = all.

Gray, C. (2004, February 11). Meetup.com working to become a force in local, state politics. *Knight Ridder Tribune*. Retrieved from http://www.highbeam.com/doc/1G1-113163315.html.

Hall, E. (2012). Year after Arab Spring, digital, social media shape region's rebirth. *Advertising Age, 83*(24), 10.

Halverson, J. R., Ruston, S. W., & Trethewey, A. (2013). Mediated martyrs of the Arab Spring: New media, civil religion, and narrative in Tunisia and Egypt. *Journal of Communication, 63*(2), 312–332. doi: 10.1111/jcom.12017.

Hampton, K., Goulet, L. S., Rainie, L., & Purcell, K. (2011). *Social networking sites and our lives*. Retrieved from http://www.pewinternet.org/2011/06/16/social-networking-sites-and-our-lives/.

Hirst, M. (2012). One tweet does not a revolution make: Technological determinism, media and social change. *Global Media Journal, 12*, 1–29.

Hong, H. (2013). Government websites and social media's influence on government-public relationships. *Public Relations Review, 39*, 346–356. doi: 10.1016/j.pubrev.2013.07.007.

Human Rights Watch (2010) Iran. Retrieved from http://www.hrw.org/en/node/87713.

Ifukor, P. (2010). "Elections" or "selections"? Blogging and twittering the Nigerian 2007 general elections. *Bulletin of Science, Technology & Society, 30*(6), 398–414. doi: 10.1177/0270467610380008.

Jenkins, H. (2006). *Convergence culture*. New York: New York University Press.

Johnston, E., & Hansen, D. (2011). Design lessons for smart governance infrastructures. In D. Ink, A. Balutis, and T. Buss (Eds.), *American governance 3.0: Rebooting the public square?* (pp. 197–212). National Academy of Public Administration.

Jungherr, A. (2014). The logic of political coverage on Twitter: Temporal dynamics and content. *Journal of Communication, 64*, 239–259. doi: 10.1111/jcom.12087.

LaMarre, H. L., & Suzuki-Lambrecht, Y. (2013). Tweeting democracy? Examining Twitter as an online public relations strategy for congressional campaigns. *Public Relations Review*, *39*(4), 360-368. doi: 10.1016/j. pubrev.2013.07.009.

Lawson-Borders, G., & Kirk, R. (2005). Blogs in campaign communication. *American Behavioral Scientist*, *49*, 548-559. doi: 10.1177/0002764205279425.

Lee, E., & Weinthall, B. (2012). The truly revolutionary social networks. In T. Manhire (Ed.), *The Arab Spring: Rebellion, revolution and a new world order* (pp. 283-285). London: Guardian Books.

Lee, J. K. (2007). The effect of the Internet on homogeneity of the media agenda: A test of the fragmentation thesis. *Journalism & Mass Communication Quarterly*, *84*, 745-760. doi: 10.1177/107769900708400406.

Levenshus, A. (2010). Online relationship management in a presidential campaign: A case study of the Obama campaign's management of its Internet-integrated grassroots effort. *Journal of Public Relations Research*, *22*(3), 313-335. doi: 10.1080/10627261003614419.

Levinson, P. (2013). *New new media*. Pearson/Penguin Academics.

Linders, D. (2012). From e-government to we-government: Defining a typology for citizen coproduction in the age of social media. *Government Information Quarterly*, *29*, 446-454. doi: 10.1016/j.giq.2012.06.003.

Lindsey, R. (2013). What the Arab Spring tells us about the future of social media in revolutionary movements. *Small Wars Journal*, *9*(7). Retrieved from http://small-warsjournal.com/jrnl/art/what-the-arab-spring-tells-us-about-the-future-of-social-media-in-revolutionary-movements.

Marks, G., & Miller, N. (1987). Ten years of research on the false-consensus effect: An empirical and theoretical review. *Psychological Bulletin*, *102*(1), 72-90. doi: 10.1037/0033-2909.102.1.72.

Mattheson, D. (2004).Weblogs and the epistemology of the news: Some trends in online journalism. *New Media & Society*, *6*, 443-468. doi: 10.1177/146144804044329.

McCafferty, D. (2011). Activism vs. slacktivism. *Communications of the ACM*, *54*(12), 17-19. doi: 10.1145/2043174.2043182.

McLuhan, M. (1964). Understanding media: The extension of man. London: Sphere Books. (van) Noort, G., Antheunis, M., & van Reijmersdal, E. (2012). Social connections and the persuasiveness of viral campaigns in social network sites: Persuasive intent as the underlying mechanism. *Journal of Marketing Communications*, *18*(1), 39–53. doi: 10.1080/13527266. 2011.620764.

Norris, D. F., & Reddick, C. G. (2011). *E-government 2011 survey*. Retrieved from http://icma.org/en/icma/knowledge network/documents/kn/Document/302947/.

Norris, D. F., & Reddick, C. G. (2013). Local e-government in the United States: Transformation or incremental change? *Public Administration Review*, *73*(1), 165–175. doi: 10.1111/j.1540-6210.2012.02647.x.

Papacharissi, Z., & de Fatima Oliveira, M. (2012). Affective news and networked publics: The rhythms of news storytelling on #Egypt. *Journal of Communication*, *62*, 266–282. doi: 10.1111/j.1460-2466.2012.01630.x.

Pew Charitable Trust (2008). *A record-breaking 46 percent of Americans have already used Internet for politics this election season*. Retrieved from http://www.pewtrusts.org/en/about/news-room/press-releases/2008/06/15/a-recordbreaking-46-of-americans-have-already-used-internet-for-politics-this-election-season.

Pew Research Journalism Project (2012a, August 15). *How the presidential candidates use the web and social media*. Retrieved from http://www.journalism.org/2012/08/15/how-presidential-candidates-use-web-and-social-media/.

Pew Research Journalism Project (2012b, August 15). *Messaging-two different strategies*. Retrieved from http://www.journalism.org/2012/08/15/messaging-two-different-strategies/.

Preston, J. (2011, February 5). Movement began with outrage and a Facebook page that gave it an outlet. *New York Times*. Retrieved from http://www.nytimes.com/2011/02/06/world/middleeast/06face.html?pagewanted=all&_r=0.

Sadeghi, L. (2012). Web 2.0. In M. Lee, G. Neeley, & K. Stewart (Eds.), *The practice of government public relations* (pp. 25–140). Boca Raton, FL: CRC Press.

Salem, F., & Mourtada, R. (2011). *Facebook usage: Factors and analysis.* Arab Social Media Report #2. Dubai: Dubai School of Government.

Schario, T., & Konisky, D. (2008). *Public confidence in government: Trust and responsiveness*. Retrieved from http://truman.missouri.edu/ipp/publications/index.asp?.

Searson, E. M., & Johnson, M. A. (2010). Transparency laws and interactive public relations: An analysis of Latin American government Web sites. *Public Relations Review*, *36*(2), 120–126. doi: 10.1016/j.pubrev.2010.03.003.

Sifry, M. (2011). *From Howard Dean to the tea party: The power of meetup.com*. Retrieved from http://www.cnn.com/2011/11/07/tech/web/meetup-2012-campaign-sifry/.

Sunstein, C. R. (2007). *Republic.Com 2.0*. Princeton, NJ: Princeton University Press.

Takaragawa, S., & Carty, V. (2012). The 2008 U.S. Presidential election and new digital technologies: Political campaigns as social movements and the significance of collective identity. *Tamara: Journal for Critical Organization Inquiry*, *10*(4), 73–89.

Tiffen, R. (1989). *News & Power*. Sydney: Allen & Unwin.

Torres, L. (2007). Citizen sourcing in the public interest. *Knowledge Management for Development Journal*, *3*(1), 134–145.

Wagner, K., & Gainous, J. (2009). Electronic grassroots: Does online campaigning work. *Journal of Legislative Studies*, *15*(4), 502–520. doi: 10.1080/13572330903302539.

Warner, B. R., & Neville-Shepard, R. M. (2011). The polarizing influence of fragmented media: Lessons from Howard Dean. *Atlantic Journal of Communication*, *19*, 201–215. doi: 10.1080/15456870.2011.606100.

Waters, R. D., & Williams, J. M. (2011). Squawking, tweeting, cooing, and hooting: Analyzing the communication patterns of government agencies on Twitter. *Journal of Public Affairs*, *11*(4), 353–363. doi: 10.1002/pa.385.

Weinberg, B., & Williams, C. (2006). The 2004 U.S. Presidential campaign: Impact of hybrid offline and online "meetup" communities. *Journal of Direct, Data and Digital Marketing Practice*, *8*(1), 46–57. doi: 10.1057/palgrave.dddmp.4340552.

White, D. M. (1950). The gatekeeper: A case study in the selection of news.

*Journalism Quarterly*, *27*, 383-390.

Woolley, J., Limperos, A., & Oliver, M. (2010). The 2008 presidential election, 2.0: A content analysis of user-generated political Facebook groups. *Mass Communication and Society*, *13*, 631-652. doi: 10.1080/15205436.2010.516864.

Wolf, G. (2004). *How the Internet invented Howard Dean*. Retrieved from http://www.wired.com/wired/archive/12.01/dean_pr.html.

Zhuo, X., Wellman, B., & Yu, J. (2011). Egypt: The first Internet revolt? *Peace Magazine*. Retrieved from http://peacemagazine.org/archive/v27n3p06.htm.

# 5

# 社交媒体隐私与安全

随着技术的迅猛发展,曾经的私密信息成了公开信息。社交网站往往会鼓励用户表露个人信息(Antheunis, Valkenburg, & Peter, 2010),比如生日、年龄、宗教信仰、政治观点、感情状态和性取向等私人信息(Gross & Acquisti, 2005)。这些个人信息可以被任何人获取,包括管理者、陌生人,甚至是心怀恶意的朋友。不少职员因为他们发布在 Facebook 上的文章、评论、照片及他们所加入的群组而被公司解雇。例如,新英格兰爱国者队(New England Patriots)[①]18 岁的啦啦队队员凯特琳·戴维斯(Caitlin Davis)因在 Facebook 上上传的照片而被解雇。2013 年,皮尤研究中心以 1 002 名 18 岁以上的成年人为对象,调查了线上匿名性、隐私和安全。1%的参与者宣称,他们曾因在线上发布信息或受到他人的线上评论而错失工作、教育机会。21%的参与者表示自己的邮箱或社交网络账号曾被其他人盗用(Rainie, Kiesler, Kang, & Madden, 2013)。

---

① 一支美式橄榄球球队,位于美国马萨诸塞州。——译者注

## 隐私悖论

许多公民担忧政府会掌握过多的个人情报,但仍会主动在社交网站上表露个人信息,研究者将此情况称为"隐私悖论"(Norberg, Horne, & Horne, 2007)。这一悖论能够从心理学的视角进行解释。人们为了融入所属的社会团体,获得人气,而表露个人信息(Christofides, Muise, & Desmarais, 2009)。因此,马威克和博伊德(Marwick & boyd, 2014)强调不应该把隐私当作个体化模式进行研究,而应该当作网络化的隐私。人们无法仅靠自身保护隐私(Marwick & boyd, 2014),因为一些朋友可能会在我们的文章下留下令人尴尬的回复,或者分享那些我们不会主动发布且有损形象的照片。社交网站用户往往在不同的群组里分享不同的信息,并依据信赖和尊重来决定内容对谁可见。

## 隐私担忧

一项近期的研究(Stiger, Burger, Bohn, & Voracek, 2013)对比了全世界的 Facebook 用户和注销账号的人,发现人们注销 Facebook 账号主要是因为隐私担忧(48%)。退出 Facebook 的人的年龄显著大于继续使用者的年龄,并且男性退出的比例更高(72%)。根据里森-鲁佩(Reason-Rupe, 2013)的隐私调查,比起 Facebook,人们更愿意相信美国国家税务局。超过 60% 的美国人表示他们"根本不"相信 Facebook 会保护他们的隐私,15% 的人表示他们对 Facebook 只有"少许"信任。相比之下,有 45% 的美国人表示他们不相信美国国家税务局会保护他们的隐私,18%的人稍微信任(Ekins, 2013)。大多数人没有意识到,尽管Facebook提供了许多隐私设置,却追踪了他们的兴趣、喜好,并根据这些信息在他们的主页推送广告。人们正成为"社交媒体隐私陷阱"的牺牲者。例如,Facebook会将你的朋友所喜欢的产品服务展示给你,从而吸引你喜欢同样的页面。

刘、格玛迪、克里希纳穆尔蒂和梅丝洛夫(Liu, Gummadi, Krishnamurthy, & Mislove, 2011)就Facebook的隐私设置,衡量了用户期望与实际情况的匹配度。他们发现实际情况与用户期望的匹配度仅为37%。换而言之,Facebook的隐私设置并没有足够的透明度。马德伊斯基、约翰逊和贝洛维(Madejski, Johnson, & Bellovin, 2011)也研究了用户的Facebook隐私设置是否与期望相匹配,发现每位参与者都表明Facebook的隐私设置或多或少地违背了他们的分享意图,并由此建议改善默认隐私设置。

为了打消人们社交媒体使用中的隐私担忧,泰勒·德罗尔和布鲁克斯·巴芬顿(Tyler Droll & Brooks Buffington)开发了匿名社交应用Yik Yak。Yik Yak能让用户搜索到方圆数英里内的"Yaks",通过这种方式,人们能够和附近的人们聊天。Yik Yak与Twitter相似,但实行了GPS定位功能和匿名制。一些人认为匿名是社交媒体的未来趋势,但也有人担心会由此引发网络暴力问题(Fye, 2014)。与其他社交媒体应用相同,Yik Yak收集了人们的信息和评论(这些评论可以转发到Facebook、Twitter、Instagram上)。为遵守法律、诉讼程序或法院指令,Yik Yak同样也会公开用户个人信息(yikyakapp.com, 2014),或向第三方分享用户信息,以用于市场调查与研究。

# 人力资源和隐私

在一项调查中,研究者询问大学生如何看待雇主查看他们的社交媒体页面这一做法。68%的学生认为这种行为并没有产生伦理问题(Clark & Roberts,2010)。这一结果也许说明人们对社交媒体作用的认识产生了代际变化。人们还未厘清社交网站上的内容是公开的还是个人的,因此用社交媒体来决定是否雇用一个人还存在困惑(Aase,2010)。由于能够公开搜索到的信息并不属于私人信息,雇主可以在招聘过程中将社交媒体上的信息列入考虑范围(Brown & Vaughn,2011)。但是,雇主能在多大程度上因雇员不恰当使用社交媒体而解雇对方则取决于国家法律。在过去几年中,一些雇主开始要求现有雇员和未来雇员提供自己的社交媒体账号与密码。在美国一些州,立法者已经引入了相关法律保护阻止雇主的这类做法。2013年,相关法律在10个州得以实施。2014年增至17个州。至2014年12月,已有28个州引入或待批相关法律。例如,路易斯安那州州长签署了《个人在线账号隐私保护法令》(*Personal Online Privacy Protection Act*),禁止雇主和教育机构索取个人信息以查看个人在线账号(NCSL,2014)。

执法机构则越来越频繁地使用社交媒体破案。有时,警方会违反 Facebook 的政策,创立虚假账号,从而搜集嫌疑人的信息(Kelly,2012)。民权组织如电子前沿基金会(Electronic Frontier Foundation)批评这种做法有损人们在线个人信息的隐私权(Kelly,2012)。根据一项关乎1 200名联邦、州和地方执法机构官员的调查,20%的参与者曾使用不同社交媒体平台辅助调查,过半数官员认为社交媒体能帮他们更快破案(LexisNexis,2012)。

在一些案件中,犯罪者往往会在社交媒体上掉以轻心,谋划犯罪计划或吹嘘自己的恶行(Kelly,2012)。

# 青少年的线上隐私和安全观点

2013年,皮尤研究中心调查了青少年社交媒体隐私问题(Madden, Lenhart, Cortesi, Gasser, Duggan, Smith & Beaton),60%的青少年Facebook用户表示已将自己的账户设置为私人账户,仅有16%的青少年设置了发布新内容时自动添加地理位置。此外,74%的青少年采取了从好友列表中删除一些人的隐私管理策略。这些数据看起来鼓舞人心,但青少年在Twitter上却并未这样做。Twitter没有Facebook的隐私设置功能,用户只能选择保护推文或公开推文。该研究同时表明,60%的青少年社交媒体使用者错误地认为Facebook不会将他们的数据分享给第三方平台。

总体而言,青少年由于并未意识到可能的后果,所以往往在线上"过多分享"信息(Walrave, Vanwesenbeeck, & Heirman, 2012),这同时也解释了为什么他们社交媒体账号的隐私设置等级较低。青少年易于在社交媒体上分享那些在成年人看来私密的信息,但那些不信任社交媒体联系人的用户通常拒绝分享私人信息,并会采取更严格的隐私设置(Patil, 2012)。在此,对社交媒体好友的低信任度显著提高了信息控制的程度(Christofides et al., 2009)。赫里斯托菲季斯等人(Christofides et al., 2009)也发现,信息表露和信息控制并非呈显著负相关关系,这与之前的其他研究和假设不同。社交媒体上的表露行为源自人们对人气的需求和表露的一般性倾向,而自尊和信任度预示了用户的信息控制度。在研究中,学生普遍担忧个人隐私问题,然而他们往往会分享其他用户也分享的个人信息。这种行为与理性行为理论（Ajzen & Fishbein,1980）相符。理性行为理论认为,人的行为取决于他/她实施行为的意图,这种意图反过来作用于他/她对行为的态度和他/她的主观规范。主观规范是指"个人对于是否实施行为所感受到的社会压力"（Ajzen,1991）。这种社会压力可能源于同侪、家庭、学校或工作场合,也可能是学生们对公开自己的生活感到有压力的原因。

在第三章,我们讨论了人们对名声的追求。追求名声是青少年在Facebook上自我表露的一个原因。在Facebook等实名环境中,人们倾向于展示、炫耀自己,而非告诉别人自身的情况（Zhao, Grasmuck, & Martin,2008）。事实上,大部分青少年更在意控制他们的父母能够看到什么（boyd,2007）。他们经常会创建不同的读者群体,有些人能看到他们的文章,另一些人则不能。青少年还会发布一些只有最亲密的朋友才能理解的内容。博伊德和马威克（boyd & Marwick,2011）引入"社交速写"（social steganography）的概念来定义这种隐藏信息并只与特定好友实现交流的行为。例如,在佩鲁切特和卡尔（Peluchette & Karl,2008）的研究中,20%的参与者表示,如果他们的雇主看到他们Facebook上的特定内容,他们会感到不适。学生最不希望雇主知晓的是与酒精饮品有关的照片和评论。一些学生不想让雇主看到朋友们某些不合时宜的玩笑与评论。就性别差异而言,女性更关心自己的社交媒体隐私,会对自身的账号页面设置更多限制（Walrave, Vanwesenbeeck, & Heirman,2012）。佩鲁切特和卡尔（2008）则发现男性更易于发布一些推销自己和有伤风化的照片评论（与性、酒精饮品相关）,而女性更可能发布一些浪漫或"可爱"的图片。

其他研究也发现隐私担忧与严格的隐私设置呈正相关（Utz &

Kramer, 2009; Nov & Wattal, 2009)。尽管当下的青少年会发布长辈认为是隐私的个人信息（如年龄、政治立场、宗教信仰、性取向），但他们也注意到将个人信息发布到网上的风险（Utz & Kramer, 2009）。利文斯通（Livingstone, 2008）认为，青少年对隐私的定义发生了变化；隐私不再是关乎特定信息是否被揭露的问题，而是关乎信息控制和哪些人了解他们的哪些个人信息的问题。青少年对父母设置访问权限就印证了这种看法。

使用社交媒体的另一个负面影响是高中生与大学生提及酒精饮料及毒品的话题，并且被公众，甚至未来的雇主看到。伊根和莫雷诺（Egan & Moreno, 2011）检验了男性本科生在 Facebook 页面上提及酒精饮品话题的情况，发现达到饮酒合法年龄的人提及酒精饮品的次数是其他人的 4.5 倍。在所有的 Facebook 个人页面上，85.3% 的人提及了酒精饮品的相关话题。伊根和莫雷诺（2011）认为，这种描述饮酒活动的行为可能会给个人和职业声誉带来负面影响。

## 传播隐私管理理论

最近，传播隐私管理理论（Petronio, 2002）被用来解释"隐私悖论"，或解释人们表达隐私担忧但仍在社交媒体网站表露个人信息的原因（第一章对这一理论展开了讨论）。彼得罗尼奥（Petronio, 2002, p.6）将隐私定义为"人们感到个人有权利从个人及集体的维度来控制个人信息"。一个人分享私人信息后，其他人成为共同拥有者，而这可能会导致隐私紊乱，或者使故意违背默认规则的有心人将理应保持私密的信息表露出来。另一种隐私管理技巧是添加好友、删除好友和屏蔽好友（Madden et al., 2013）。根据一项研究,74%的青少年社交媒体用户曾从他们的好友列表里删除特定人。这种行为表明,青少年对于向谁分享信息这一问题,采取了积极主动的想法。

　　大学教授和教育人士每天处理隐私辩证法问题,他们需要思考在社交媒体上,究竟应向学生展示或隐藏哪些内容。我们会听说有些教师在社交网络上加学生为好友,但目前没有学者研究过教师的隐私担忧问题。一些研究者(Cain & Fink, 2010)认为过度暴露彼此的私人生活可能会对教师产生负面影响,因为学生通常不希望教师出现在他们的社交网络上(Hewitt & Forte, 2006)。据报道,由于在社交网络上公布某些内容,一些初高中、大学教师被免职。根据美国大学教授联盟(American Association of University Professors, AAUP)拟定的伦理原则和标准,大学教职人员必须"避免与学生陷入双重关系,避免因身兼多重角色而引起责任冲突、角色混乱、期望含糊,进而致使教职人员涉入不可兼容的角色和互相冲突的责任之中"(ACPA Ethics Code, 2006)。

# 参 考 文 献

Aase, S. (2010). Toward e-professionalism: thinking through the implications of navigating the digital world. *Journal of the American Dietetic Association*, *110*(10), 1442–1447. doi: 10.1016/j.jada.2010.08.020.

ACPA Ethics Code (2006). *Statement of ethical principles and standards*. Retrieved from http://www.myacpa.org/au/documents/EthicsStatement.pdf.

Ajzen, I. (1991). The theory of planned behavior. *Organization Behavior and Human Decision Processes*, *50*, 179–211. doi: 10.1016/0749-5978(91)90020-T.

Ajzen, I., & Fishbein, M. (1980). *Understanding attitudes and predicting social behavior.* Englewood Cliffs, NJ: Prentice-Hall.

Antheunis, M. L., Valkenburg, P. M., & Peter, J. (2010). Getting acquainted through social network sites: Testing a model of online uncertainty reduction and social attraction. *Computers in Human Behavior*, 26, 100–109. doi: 10.1016/j.chb.2009.07.005.

boyd, d. m. (2007). Why youth (heart) social network sites: The role of networked publics in teenage social life. In Buckingham, D. (Ed.), *McArthur Foundation series on digital learning—Youth, identity, and digital media volume*. Cambridge, MA: MIT Press.

boyd, d. m., & Marwick, A. (2011) Social privacy in networked publics: teens' attitudes, practices, and strategies. Paper presented at the Oxford Internet Institute Decade in Internet Time Symposium.

Brown, V. R., & Vaughn, E. D. (2011).The writing on the (Facebook) wall: The use of social networking sites in hiring decisions. *Journal of Business Psychology*, 26, 219–225. doi: 10.1007/s10869-011-9221-x.

Cain, J., & Fink, J. L. (2010). Legal and ethical issues regarding social media and pharmacy education. *American Journal of Pharmaceutical Education*, 74(10), 1–7. doi: 10.5688/aj7410184.

Christofides, E., Muise, A., & Desmarais, S. (2009). Information disclosure and control on Facebook: Are they two sides of the same coin or two different processes? *CyberPsychology & Behavior*, 12, 341–345. doi: 10.1089/cpb.2008.0226.

Clark, L., & Roberts, S. (2010). Employer's use of social networking sites: A socially irresponsible practice. *Journal of Business Ethics*, 95(4), 507–525. doi: 10.1007/s10551-010-0436-y.

Egan, K., & Moreno, M. (2011). Alcohol references on undergraduate males' Facebook profiles. *American Journal of Men's Health*, 5, 413–420. doi: 10.1177/1557988310394341.

Ekins, E. (2013). *Poll: On privacy, IRS, and NSA deemed more trustworthy than Facebook and Google.* Retrieved from http://reason.com/poll/2013/09/27/poll-on-privacy-irs-and-nsa-deemed-more.

Fye, S. (2014). Yik Yak: Why it exists. *The Atlas Business Journal.* Retrieved from http://atlasbusinessjournal. org /yik-yak-greater-implications-upon-society/.

Gross, R., & Acquisti, A. (2005). Information revelation and privacy in online social networks. *Proceedings of the 2005 ACM workshop on privacy in the electronic society*, 71-80.

Hewitt, A., & Forte, A. (2006). *Crossing boundaries: Identity management and student / faculty relationships on Facebook*. Poster presented at the Computer Supported Cooperative Work (CSCW).

Kelly, H. (2012, August 30). *Police embrace social media as crime-fighting tool*. Retrieved from http://www.cnn.com/2012/08/30/tech/social-media/fighting-crime-social-me-dia/.

LexisNexis (2012). *Role of social media in law enforcement significant and growing*. Retrieved from http://www.lexisnexis.com/en-us/about-us/media/press-release.page?id=1342623085481181.

Liu, Y., Gummadi, K., Krishnamurthy, B., & Mislove, A. (2011). Analyzing Facebook privacy settings: User expectations vs. reality. In *Proceedings of the 2011 ACM SIGCOMM conference on Internet measurement conference*, 61-70.

Livingstone, S. (2008). Taking risky opportunities in youthful content creation: Teenagers' use of social networking sites for intimacy, privacy, and self-expression. *New Media & Society*, *10*, 339-411. doi: 10.1177/1461444808089415.

Madden, M., Lenhart, A., Cortesi, S., Gasser, U., Duggan, M., Smith, A., & Beaton, M. (2013, May 21). *Teens, social media, and privacy*. Retrieved November 1, 2014, from http://www.pewinternet.org/2013/05/21/teens-social-media-and-privacy/#.

Madejski, M., Johnson, M., & Bellovin, S. (2011). The failure of online social network privacy settings. *Columbia University Computer Science Technical Reports*. Retrieved from http://academiccommons.columbia.edu/catalog/ac:135406.

Marwick, A. E., & boyd, d. (2014). Networked privacy: How teenagers negotiate context in social media. *New Media & Society*, *16*(7), 1051-1067. doi: 10.1177/1461444814543995.

NCSL (2014). Employer access to social media usernames and passwords.

Retrieved from http://www.ncsl.org/research/telecommunications-and-information-technology/employer-access-to-social-media-passwords-2013.aspx.

Norberg, P. A., Horne, D. R., & Horne, D. A. (2007). The privacy paradox: Personal information disclosure intentions versus behaviors. *Journal of Consumer Affairs*, *41*, 100-126. doi: 10.1111/j.1745-6606.2006.00070.x.

Nov, O., & Wattal, S. (2009). *Social computing privacy concerns: antecedents and effects*. Paper presented at 27th International Conference on Human Factors in Computing Systems.

Patil, S. (2012). *Will you be my friend?: Responses to friendship requests from strangers*. Proceedings of the 2012 iConference.

Peluchette, J., & Karl, K. (2008). Social networking profiles: An examination of student attitudes regarding use and appropriateness of content. *CyberPsychology & Behavior*, *11*, 95-7. doi: 10.1089/cpb.2007.9927.

Petronio, S. (2002). *Boundaries of privacy: Dialectics of disclosure*. New York: State University of New York Press.

Rainie, L., Kiesler, S., Kang, R., & Madden, M. (2013). Anonymity, privacy, and security online. *Pew Research Internet Project*. Retrieved from http://www.pewinternet.org/2013/09/05/anonymity-privacy-and-security-online/.

Stieger, S., Burger, C., Bohn, M., & Voracek, M. (2013). Who commits virtual identity suicide? Differences in privacy concerns, Internet addiction, and personality between Facebook users and quitters. *Cyberpsychology, Behavior, and Social Networking*, *16*, 629-634. doi: 10.1089/cyber.2012.0323.

Utz, S., & Kramer, N. (2009). The privacy paradox on social network sites revisited: The role of individual characteristics and group norms. *Cyberpsychology: Journal of Psychosocial Research on Cyberspace*, *3*(2), article 1.

Walrave, M., Vanwesenbeeck, I., & Heirman, W. (2012). Connecting and protecting? Comparing predictors of self-disclosure and privacy settings use between adolescents and adults. *Cyberpsychology: Journal of Psychosocial Research on Cyberspace*, *6*(1), article 3.

YikYak (2014). *Privacy*. Retrieved from http://www.yikyakapp.com/privacy/.

Zhao, S., Grasmuck, S., & Martin, J. (2008). Identity construction on Facebook: Digital empowerment in anchored relationships. *Computers in Human Behavior, 24*, 1816-1836. doi: 10.1016/j.chb.2008.02.012.

# 6

# 社交媒体与教育

传统的课堂环境包括教师和学生。教师的任务是传授课程内容,其授课过程通常是单向的。新传播技术则为教学过程中的互动式学习提供了机会,教师不再是课堂的中心。比如,社交媒体使得教师和学生在课堂内外的互动成为可能。但一些研究指出,教师们并没有做好在课堂中使用社交媒体的准备。许多教师仍担心存在网络安全、隐私、考试作弊(Sharples, Graber, Harrison, & Logant, 2008)及工作时间延长(Lepi, 2013)等问题。尽管大部分学生热衷于在课堂上使用社交媒体,但教师们却采取慢慢来的态度(Queirolo, 2009)。而为了让社交媒体和教学相结合,教师不仅要转变教学方式,还应学会将社交媒体作为教学工具来使用(Keengwe, Kidd, & Kyei-Blankson, 2009; Long, 2009)。

本章探讨了教育领域中社交媒体应用的优势和挑战;展示了支持在大学课堂上使用博客、YouTube 和 Twitter 的实验研究结果;进一步阐释了 Facebook 上师生关系的动态变化;最后以讨论 e-learning 和社交媒体在中小学教育(K-12)中的应用收尾。

# 教育领域中社交媒体应用的优势和挑战

通过质疑学校、教师、学生和学习中的先入之见,社交媒体多方面地挑战了传统教育(Bartow,2014;Condie & Livingston,2007)。巴托(Bartow,2014)指出,在学校结构和组织中,社交媒体技术正呈现出教育的、伦理的、经济的及革新的变化。在教育层面,自主学习是变化之一。用户可以自主地选择何时、何地、以何种方式接受谁的教育(Bartow,2014)。同时,社交媒体还实现了思想交流的开放与自由。在伦理层面,将数字技术纳入公共教育有助于缩小数字鸿沟。在经济层面,在线教育可以低成本地覆盖大量学生。最后,在革新层面,社交媒体则实现了知识获取的民主化(Bartow,2014)。

下面列出了在教育中使用社交媒体的优势:

- 以学生为中心的课程增多(Greenhow,2011)。
- 学生之间、师生之间的互动增加(Hoffman,2009)。
- 个性化选择和定制(Hoffman,2009)。
- 用户生产内容,允许使用维基和博客进行协作(Gikas & Grant,2013;Lemoine & Richardson,2013)。
- 可访问性和实用性得到改进;越来越多的成年学生参与到学习中(Bjerede, Atkins, & Dede,2010)。
- 学生参与度和互动水平得到提高(Nicolini, Mengis, & Swan,2012)。
- 害羞的学生在这种间接式的环境中参与度更高(Van Merriënboer & Stoyanov,2008)。
- 学生可以使用 YouTube 分享视频作业或用 Twitter 来

关注某些观点（Junco, Heiberger, & Loken, 2011; Shih & Waugh, 2011）。

● 为创作及获得他人的支持提供了机会（Greenhow & Robelia, 2009）。

## 利用社交媒体招收和留住学生

如今，很多大学和学校都开始利用社交媒体来招收新生、与在校生分享信息。高中生们通过社交媒体了解他们理想中的学院和大学。根据 2013 年的《社会招生报告》（*2013 Social Admissions Report*）显示，当年有 75% 的高中毕业生在选择大学时参考了社交媒体上的信息。在学生们用来查找相关大学信息的社交媒体中，Facebook 名列第一，紧随其后的是 YouTube、Twitter 和 Instagram。尽管 75% 的学生将社交媒体作为消息源，但只有 49% 的人喜欢或关注了相关学校的社交媒体账号。有趣的是，学生最想与之互动的三种人是：（1）刚入校学生；（2）招生顾问；（3）其他新生。针对大学如何改善其在社交媒体上的展示，相关建议包括：发布更多照片、与奖学金和实习有关的信息、招生信息及特色课程视频。一项来自升学咨询公司（Zinch）和致力于学生在线社交的公司（Inigral）的调查显示，高中生希望能与高等院校的社交媒体网站有更多的互动。如今，进入大学的是伴随着数字互动媒体成长起来的一代，因此，他们对文本数据的依赖程度较低。尽管 Facebook 拥有最多的日活跃用户量，但事实上，大多数高中生视 Instagram 为最受欢迎的社交网站（Thompson, 2014）。这很大程度上是因为 Instagram 是一个以分享照片和视频为主的网站。

社交媒体不仅可以帮助学校招生，还可以帮助他们留住学生。麦克卢尔（McClure，2013）举例指出，如果有学生在社交媒体上吐槽某节课，在线社群的管理人员和其他学校人员就能够在问题恶化前找到这些学生并解决问题。过去，学生们会当面向他们的同龄人抱怨，或者打电话给家人，但现在，他们会在社交媒体上广播他们所关心的事情（McClure，2013）。因此，大学需要创建一个社交媒体网站，在那里，学生们可以和那些选同一门课，或有相似经历的同学进行交流。

教育学者认为，同学是大学生生活最重要的影响因素，也是教育成果——包括耐心和承诺——的关键预测变量（Astin，1993；Broome，Croke，Staton，& Zachritz，2012）。社交媒体可以让学生们在线分享和交流想法。这对大学一年级新生而言，通常尤为重要，新生们会通过社交媒体努力结交新朋友以适应大学生活（Galindo，Bogran Meling，Mundy，& Kupczynski，2012）。赫南德斯、纽曼和洛佩兹（Hernandez，Newman，& Lopez，2014）建议各所大学应当更有效地使用Facebook，包括给与其目标和受众相关的页面点赞、根据受众的时间安排发布消息的时间，以及标记相关页面从而增加影响力。学校还应利用线下渠道宣传Facebook页面，使用不同类型的帖子来保持页面动态，将Facebook与其他渠道相连，以及保持创造力。赫南德斯等人（2014）还建议邀请社区学院或公立大学即将入学的新生，给该校的Facebook页面点赞、转发，学校则可以以礼品券作为回馈。赫南德斯等人（2014）还提出，大学应该在社交媒体网站上提供一些有助于学生获得成功的建议，并启用实时聊天功能，让学生能够与教师、辅导员和管理员进行实时沟通。此外，学校需要利用Facebook Analytics、Google Analytics和YouTube Analytics来评估自身的社交媒体使用行为，并保持良好的使用评比程度。最后,学校也应该密切关注粉丝们和关注者的动态。

校领导还应该鼓励学生利用社交媒体来发布他们在校所获得的成就。根据麦克卢尔(2013)的研究,这样有助于提高学校的知名度。拉米格(Ramig, 2014)则强调,学校千万不能屏蔽那些协作网站,比如博客、维基和相关社交网络。相反,教师们应学会如何使用它们。学生们则需学会如何批判地评估互联网上的信息,如何安全上网,如何适当地使用不同的网站。

## 课堂中的社交媒体

巴布森调查研究集团(Babson Survey Research Group)和皮尔森(Pearson)以8 000名高等教育机构的教师为调研对象,了解教师们如何使用社交媒体。莱皮(Lepi, 2013)的报告显示,教师对社交媒体的个人化使用水平(70%)反映了总体情况。只有41%的教师在课堂上使用社交媒体。陈和布莱尔(Chen & Bryer, 2012)分析了美国公共行政学科教师的社交媒体使用情况。在他们的研究中,教师们认为使用社交媒体的非正式学习可以融入正式的学习环境中。例如,他们相信,通过使用图片、视频和音频,可以锻炼学生的创造力。LinkedIn被视为学生联系校友和未来雇主的重要工具。一些老师已开始在教学中使用YouTube上的视频和维基上的案例研究,另一些老师则将博客作为课堂讨论的一部分。但是,教师们对于网络安全和隐私的担忧仍然存在。有些教师因为担心个人信息会泄露,或担心学生会在网上发布一些不合适的内容,而不愿在社交媒体上和学生成为朋友。对于很多年长的教师来说,最大的问题则是时间和技术障碍。陈和布莱尔(2012)指出,教师们应当尝试在课堂中使用社交媒体,而学校则应为这种尝试提供方便。

尽管社交媒体平台并不是专门为课堂使用而设计,但是在课堂外,教授们正用它们与学生进行交流并创造各种学习机会。学生们也会通过这些社交网站分享课程资料。在社交媒体发展初期(2000—2010),教授和学生用博客或在线日志来评价和讨论课程资料(Anderson,2007)。直至2011年,微博取代了传统博客。现在,Twitter是最受欢迎的微博网站,师生们用它来标记、转发相关链接,并就课程内容进行提问(McEwan,2012)。尽管并非所有教师都会在课堂上使用Twitter,但研究表明,微博的使用不仅增加了学生的课堂参与度,还提高了学生的成绩(Junco,2011;Schirmer,2011)。课后,很多学生还会继续利用社交媒体寻求学习指导,或与教授们保持联系。目前,LinkedIn是维系这种学术型联系最理想的场所。另外,学生们也可以加入Facebook小组同教师进行交流(Helvie-Mason,2011)。根据刘、考克、基尼和奥尔(Liu,Kalk,Kinney,& Orr,2009)的研究,高校中最常用的社交媒体技术是博客、播客、社交网络和虚拟环境。首先,我们对作为教育工具的博客展开讨论。

## 博客

几位学者(例如,Churchill,2009;Harrison,2011;Sim & Hew,2010;Deng & Yuen,2011)研究了在大学课堂上作为教育工具的博客。在梳理了24篇文献的基础上,西姆和休(Sim & Hew,2010)总结了博客的六大用途:(1)一种用来收集或汇报课程相关信息的学习型日志;(2)记录个人生活;(3)表达情感的途径;(4)与他人进行社交互动的交流工具;(5)同侪评估的工具；(6)发布作业的管理工具。

西姆和休(2010)还梳理了有关写博客对人们学习、思考能力的影响的文献。虽然大多数研究都是自陈报告，但研究者发现，使用博客有助于学生学习。对一些学生而言，博客提供了一个思考和发表意见的空间(Xie，Ke，& Sharma，2008)。对那些在课堂讨论中保持沉默的学生而言，博客是他们分享想法的地方。博客还可以作为数字档案记录学生的作业(Lin & Chang，2010)。杨和张(Yang & Chang，2011)在本科生和研究生中进行实验，研究他们使用博客的情况，以及当学生使用博客评论他人作业时和阅读并评论他人作业时，学生对于在线的同侪互动、同侪学习的态度及向同侪学习的动机的差异。结果表明，与单独使用博客相比，高校中博客的交互式使用，与学生对在线同侪互动的积极态度及更高的学业成就关联性更强。

大学生也会从这种与课堂主题相关的博客中获益(Churchill，2009；Harrison，2011)。博客的使用能增强学生对课程材料的参与度，从而令他们感受到自身意见的重要性。当学生有机会思考所发布的内容、评论同学发布的内容时，博客的作用才能被视为有效(Churchill，2009)。而电子邮件、pdf 讲稿和在线讨论论坛等较老的技术，也可以与 web 2.0 时代的新技术结合使用，如学生博客、课堂的维基项目、Twitter 上组织的讨论或者 YouTube 上的演讲视频(Friedman & Friedman，2013)。弗里德曼和弗里德曼(Friedman & Friedman，2013)认为，最好的解决方式是将传统的课堂学习和在线学习相结合，从而建设多样化的课程。

## Twitter

如果使用得当，Twitter 会是一种十分智能化的教学工具。乔内尔、艾尔斯和比森(Journell, Ayers, & Beeson, 2014)报告了一项关于 2012 年美国总统大选期间,高中公民学课程上的 Twitter 使用情况的调研结果。他们发现,Twitter 使学生们能以他们喜欢的方式参与到学术讨论中。卡顿(Carton, 2014)探讨了 Twitter 给教师职业发展带来的优势。不同于 Facebook,大多数人不会在 Twitter 上晒娃、晒美食。相反,他们会关注和教育相关的推文。Twitter 的另一大特点是,人们可以随兴趣的改变,轻松取消对他人的关注。人们转发的推文所包含的链接也可能会给他人带来帮助。在乔内尔等人(2014)的研究中,学生们于学期初创建自己的 Twitter 账户。老师要求学生关注美国两大党的总统候选人,并给学生们设置了专属标签(#chscivics),供他们在每个相关事件中发推文时使用。标签不仅能帮助学生看到彼此的推文,还可以让学生同全美的课堂联系起来。乔内尔等人(2014)认为这是一个将课程与现实社会相联系的典型案例(Newmann & Wehlage, 1993)。

洪科等人(Junco et al., 2011)发现,Twitter 和教学的结合能给学生的课堂参与和成绩带来积极的作用。结果表明,学生之间可以相互激励、相互合作,他们也乐于通过 Twitter 上组建的学习小组对服务项目进行协作学习。Twitter 还可以帮助那些课堂参与消极的学生融入在线讨论中。在笔者的社交媒体课堂上,学生们会将与同期课堂阅读材料有关的讨论问题发布到 Twitter 上。笔者注意到,学生们会关注其他人发布的问题,并基于此进行追问。在笔者看来,当学生被要求思考实际情况的应用性问题及自身经历与研究发现相冲突的挑战性问题时,这种课堂上 Twitter 的应用能够帮助学生批判性和创造性地思考。

因为 Twitter 并不仅是为教育目的而生的,所以这种社交媒介的使用必然存在一定的局限性,如评论不恰当、网络欺凌(Journell et al., 2014)。乔内尔等人(2014)建议,就像他们为传统课堂讨论制定规则一样,教师们需要为发布推文建立规则。此外,在他们的研究中,学生们似乎并不相互回应推文,他们更像是"谈论彼此而不是相互交谈"(p.66)。

## Facebook

一般而言,Twitter和博客仅被用于学习过程,而Facebook则是一种能够让学生和老师建立更多课堂外非正式关系的社交网络。然而对很多教师而言,这种关系具有争议性,因为在某些情况下,学生认为教师不应该出现在社交媒体上来"监视"他们的生活,老师们也会觉得学生无权进入他们的私人生活(Cain, Scott, & Akers, 2009; Hewitt & Forte, 2006)。对于在社交媒体上添加学生为朋友这一做法,很多老师都表达了伦理方面的担忧(ACPA Ethics Code, 2006)。他们尤其担心(师生)角色混乱、责任冲突及可能造成无法接受的双重关系的矛盾角色。一些学者(McEwan, 2012)主张,教师不应该在社交媒体上主动添加学生,而是要等学生来添加他们。另一些学者(如Jones, 2011)则强调,教师不应该与学生在线上闲聊,而应以教学为目的。但是,有学者(如Schwartz, 2009)提出,Facebook上教师与学生之间合乎师生关系准则的朋友关系能够让学生感受到与教师之间的关联。

如果运用得当,社交网站也可以促进课堂内外的学习。美国最高法院也秉持如下观点,即教师在社交网络上与学生的交流应受到学术自由的保护,也的确有不少研究支持这一观点。在不同的研究中,那些在Facebook上与教师成为好友的学生,更愿意与教师展开线下的互动（Sturgeon & Walker，2009）；如果教师在Facebook进行自我表露，那么学生可能会期待更活跃的课堂气氛（Mazer，Murphy，& Simonds，2007），也会带来更高的学术成就（Pascarella，1985）和更多的幸福感（Roorda，Koomen，Spilt，& Oort，2011）。普齐奥（Puzio，2013）总结道，在Facebook上，学生也许会和老师讨论校园中发生的其他问题，比如校园欺凌或抑郁。因为对学生来说，在网络上讨论这些问题要比面对面讨论更容易。"社交网站有潜力成为加强教育、沟通和学习的强有力的工具。"（Puzio，2013，p.1120）谢尔顿（Sheldon，2014）通过一项研究探讨是什么促使教师和学生在Facebook上成为好友。大多数学生添加老师为好友，主要是便于私下更好地了解他们，这一观点支持了迪威尼尔和霍谢克（DiVerniero & Hosek，2011）的研究。他们的研究发现，通过浏览老师的线上个人资料，学生们可以将老师视为"真实的人"和"朋友"。然而，老师们则对在Facebook上和学生成为好友有所顾虑，担心其他重要的人可能会对这种行为有负面看法。许多教师都提到，他们可以在学生毕业以后添加其为好友，但前提是要学生主动在线上发送好友申请。

除美国外的其他国家的学者（Draskovic，Caic，& Kustrak，2013；Baran，2010）还发现，与教师相比，更多学生认为师生通过Facebook开展社交活动是合情合理的，其中的好处包括：可以迅速地从老师和研究生助教那里得到回复，讨论课程内容，以及回答学生可能会遇到的问题（Draskovic et al.，2013）。

## 社交媒体和在线学习

教师和学生为创造新的教学方式所做的努力，导致了诸如elearning 2.0(网络学习 2.0)(Downes, 2005)和 pedagogy 2.0(教育学 2.0)等概念的出现。其特征包括开放、协作、社交网络、社会存在、用户生产内容(Dabbagh & Kitsantas, 2012)。在教育和网络学习中运用社交媒体,生成了 PLE(personal learning environment,个人学习环境)的概念。PLE 被定义为"工具、社区和服务,它们构成了个人化教育平台。学习者利用这个平台指导自身学习,并追求教育目标"(EDUCAUSE Learning Initiative [ELI], 2009, p.1)。传统的学习管理系统(learning management systems, LMS)一直受制于大学,相比之下,PLE 组织化程度低,且允许同侪间非正式的学习和联系(Dabbagh & Kitsantas, 2012)。PLE 包括 Facebook、YouTube、博客和维基等社交媒体。马丁代尔和道迪(Martindale & Dowdy, 2010)认为,社交媒体提供了类似于面对面午餐讨论和研讨会的机会。

美国国家科学基金会网络学习工作组将网络学习定义为:以网络计算和通信技术为中介的学习(National Science Foundation, 2008)。包括社交媒体在内的很多新兴的技术形式(Lemoine & Richardson, 2013)均可以帮助学生在实践中学习(Jones, Morales, & Knezek, 2005)。如今,教师不再是知识的唯一来源,更多的是在学生的学习中扮演引导者的角色(Mishra, Lemoine, Campbell, Mense, & Richardson, 2013)。

参加在线课程的学生越来越多,30% 的大学生至少上过一门网络课程(Allen & Seaman, 2010;引自 Friedman & Friedman, 2013),并且许多高校乐于开设在线课程,因为这样可以节省资金。弗里德曼和弗里德曼(2013)认为,使用社交媒体能够让作业变得有趣又实用。他们指出,与传统课堂相比,在线学习有压倒性的优势,它能让学生更好地表现自己。但是,在线学习要求学生成为学习环境的积极参与者与创造者,并且需要互相沟通。

一些人（如 Carey，2012）认为，MOOCs（Massive Open Online Courses，大规模开放在线课程）将很快改变高等教育的未来。MOOC 是无限的，通过互联网开放接入。2011 年，美国斯坦福大学提供了三门 MOOC 课程。塞巴斯蒂安·特龙（Sebastian Thrun）的《人工智能入门》（Introduction to Artificial Intelligence）课程，注册人数很快达到了几百人。大多数 MOOC 课程使用视频去呈现传统的授课内容。其中有一类课程设计是翻转课堂法。在翻转课堂中，学生在家观看在线课程，在教室进行实践。

## 社交媒体与集体智慧

一些学者（如 O'Reilly & Battelle，2009）认为，博客、维基和社交网站这类社交媒体技术，便利了由社区共同创造且服务于社区的、代表"集体智慧"的内容。汤普森、格雷和金（Thompson，Gray，& Kim，2014）基于实证研究检验了社交媒体能够促进集体智慧的假设。在由 20 位学生组成的焦点小组中，并未发现关于"集体智慧"的证据。那些利用不同社交媒体技术来完成作业的学生，依旧使用第一人称单数代词（"我"）来描述他们的学习。

另一项研究采用了由大学生所构成的大样本（N = 2 368），探讨了 Facebook 的使用与学生参与之间的关系（Junco，2012）。参与被定义为"学生投入在学术活动中的生理精力和心理精力的总和"（Astin，1984，p.297）。结果显示，学生花费在 Facebook 上的时间与学生参与程度呈负相关关系。而一些交际活动（评论、创造或回复事件）与参与程度呈正相关，非交际活动（玩游戏）则是负相关。Facebook 上的交际活动与花费在课外活动中的时间呈正相关，而非交际活动则与学生花费在课外活动上的时间呈负相关。

洪科(2012)认为学生使用 Facebook 的方式多种多样,可以借此来正向或反向预测他们的学术成果。

## 社交媒体在中小学教育中的应用

在中小学教育(K-12)中有效地使用社交媒体和在高等教育中的使用相比可能有所不同。赫夫曼(Huffman,2013)认为,老师和学生都应该接受培训,学会在教室中使用(社交媒体)技术。她认为教师不可以用个人社交账号与学生或家长保持联系。她在文章中引用了格里诺(Greenhow,2009)的建议,即在传统教室环境中,老师不会和学生分享日常生活中的琐碎和辛苦,那么在虚拟环境中也应如此。

马奥(Mao,2014)通过混合方法设计实验,对高中生如何使用社交媒体、他们对这些技术的态度和看法以及如何在教育中运用社交媒体进行了探究。马奥(2014)发现,大部分高中生利用社交媒体进行娱乐、与家人和朋友联系、分享图片和视频。与学校工作或学习相关的社交媒体使用行为则较少。学生们表示在教育中使用社交媒体十分有趣,但却很少有人这样做。大多数教师会在课堂上播放 YouTube 上的视频,但也仅限于此。这让学生有挫败感,他们认为在课堂上,社交媒体并没有物尽其用。一些学生提到,播放 YouTube 上的视频没什么实质性意义,这种行为只是教师上课的替代品。他们建议教师以更具互动性、有深度的方式来使用社交媒体(如可以利用 Skype 采访大屠杀的幸存者)。学生们则可以利用社交媒体在做作业时及时得到同学的帮助,或利用社交媒体做文献及材料的探索研究,以及获得教材的补充性资源(Mao,2014)。

# 参 考 文 献

ACPA Ethics Code (2006). *Statement of ethical principles and standards.* Retrieved from http://www.myacpa.org/au/documents/EthicsStatement.pdf.

Allen, I. E., & Seaman, J. (2010, November). Class differences: Online education in the United States, 2010. *Sloan Consortium.* Retrieved from http://sloanconsortium.org/publications/survey/class_differences.

Anderson, P. (2007). What is Web 2.0? Ideas, technologies, and implications for education. *JISC Report.* Retrieved from www.jisc.ac.uk/media/documents/techwatch/tsw0701b.pdf.

Astin, A. (1993). *What matters in college? Four critical years revisited.* Jossey-Bass.

Astin, A. (1984). Student involvement: A developmental theory for higher education. *Journal of College Student Personnel*, 25(4), 297–308.

Baran, B. (2010). Facebook as a formal instructional environment. *British Journal of Educational Technology*, 41, 146–149. doi: 10.1111/j.1467-8535.2010.01115.x.

Bartow, S. M. (2014). Teaching with social media: Disrupting present day public education. *Educational Studies: A Journal of the American Educational Studies Association*, 50, 36–64. doi: 10.1080/00131946.2013.866954.

Bjerede, M., Atkins, K., & Dede, C. (2010). Ubiquitous mobile technologies and the transformation of schooling. *Educational Technology*, 50(2), 3–7.

Broome, R., Croke, B., Staton, M., & Zachritz, H. (2012). The social side of student retention. *Inigral Insights.* Retrieved from http://www.10000degrees.org/wp-content/uploads/2012/11/The_Social_Side_of_Student_Retention.pdf.

Cain, J., Scott, D. R., & Akers, P. (2009). Pharmacy students' Facebook activity and opinions regarding accountability and e-professionalism. *American Journal of Pharmaceutical Education*, 73(6), 1–6.

Carey, K. (2012, September 7). Into the future with MOOC's. *Chronicle of Higher Education*, *59*, A136.

Carton, M. T. (2014). Twitter PD. *Illinois Reading Council Journal*, *42*, 25-27.

Chen, B., & Bryer, T. (2012). Investigating instructional strategies for using social media in formal and informal learning. *The International Review of Research in Open and Distance Learning*, *13*, 87-104.

Churchill, D. (2009). Educational applications of Web 2.0: Using blogs to support teaching and learning. *British Journal of Educational Technology*, *40*, 179-183.

Condie, R., & Livingston, K. (2007). Blending online learning with traditional approaches: Changing practices. *British Journal of Educational Technology*, *38*, 337-348.

Dabbagh, N., & Kitsantas, A. (2012). Personal learning environments, social media, and self-regulated learning: A natural formula for connecting formal and informal learning. *Internet and Higher Education*, *15*, 3-8. doi: 10.1016/j.iheduc.2011.06.002.

Deng, L., & Yuen, A. H. K. (2011). Towards a framework for educational affordances of blogs. *Computers & Education*, *56*(2), 441-451. doi: 10.1016/j.compedu.2010.09.005.

DiVerniero, R. A., & Hosek, A. M. (2011). Students' perceptions and communicative management of instructors' online self-disclosure. *Communication Quarterly*, *59*, 428-449. doi: 10.1080/01463373.2011.597275.

Downes S. (2005). E-learning 2.0. *eLearn magazine: Education and technology in perspective*. Retrieved from http://www.elearnmag.org/subpage.cfm?section=articles&article=29-1.

Draskovic, N., Caic, M., & Kustrak, A. (2013). Croatian perspective(s) on the lecturer-student interaction through social media. *International Journal of Management Cases*, *15*(4), 331-339.

EDUCAUSE Learning Initiative (2009). *Seven things you should know about personal learning environments*. Retrieved from http://www.educause.edu/library/resources/7-things-you-should-know-about-personal-learning-environments.

Friedman, L., & Friedman, H. H. (2013). Using social media technologies to enhance online learning. *Journal of Educators Online*, *10*(1). Retrieved from http://files.eric.ed.gov/fulltext/EJ1004891.pdf.

Galindo, A. M., Bogran Meling, V., Mundy, M., & Kupczynski, L. (2012). Social media and retention: The administrative perspective at Hispanic-serving institutions of higher education. *Journal of Studies in Education*, *2*, 103–115. doi: 10.5296/jse.v2i3.1809.

Gikas, J., & Grant, M. M. (2013). Mobile computing devices in higher education: Student perspectives on learning with cellphones, smartphones & social media. *Internet & Higher Education*, *19*, 18–26. doi: 10.1016/j.iheduc.2013.06.002.

Greenhow, C. (2009). Social scholarship: Apply social networking technologies to research practices. *Knowledge Quest*, *37*(4), 42–48.

Greenhow, C. (2011.) Online social networks and learning. *On the Horizon*, *19*(1), 4–12.

Greenhow, C., & Robelia, B. (2009). Old communication, new literacies: Social network sites as social learning resources. *Journal of Computer-Mediated Communication*, *14*, 1130–1161. doi: 10.1111/j.1083–6101.2009.01484.x.

Harrison, D. (2011). *Can blogging make a difference?: Campus Technology*. Retrieved from http://campustechnology.com/articles/2011/01/12/can-blogging-make-a-difference.aspx.

Helvie-Mason, L. (2011). Facebook, 'friending,' and faculty-student communication. In C. Wankel (Ed.), *Teaching arts and sciences with the new social media: Cutting-edge technologies in higher education* (vol. 3). Bingley, U.K.: Emerald Group.

Hernandez, C., Newman, P., & Lopez, R. (2014). *Facebook me: Using social media to promote college retention*. Retrieved from http://media.collegeboard.com/digitalServices/pdf/diversity/2014/facebook-me-using-social-media-promote-college-retention.pdf.

Hewitt, A., & Forte, A. (2006). *Crossing boundaries: Identity management and student/faculty relationships on Facebook*. Poster presented at the

Computer Supported Cooperative Work (CSCW).

Hoffman, E. (2009). Social media and learning environments: Shifting perspectives on the locus of control. *In Education: Exploring Our Connective Educational Landscape*, *15*(2), 23-38. Retrieved from http://ined.uregina.ca/ineducation/article/view/54/0.

Huffman, S. (2013). Benefits and pitfalls: Simple guidelines for the use of social networking tools in K-12 education. *Education*, *134*(2), 154-160.

Jones, A. (2011). How Twitter saved my literature class: A case study with discussion. In C. Wankel (Ed.), *Teaching arts and sciences with the new social media: Cutting-edge technologies in higher education* (vol. 3). Bingley, U.K.: Emerald Group.

Jones, J. G., Morales, C, & Knezek, G. A. (2005). 3-Dimensional online learning environments: examining attitudes toward information technology between students in Internet-based 3-dimensional and face-to-face classroom instruction. *Educational Media International*, *42*(3), 219-236.

Journell, W., Ayers, C. A., & Beeson, M. W. (2014). Tweeting in the classroom. *Phi Delta Kappan*, *95*(5), 63.

Junco, R. (2011). *Twitter to improve college student engagement*. Paper presented at SXSW Interactive, Austin, Texas, 2011.

Junco, R. (2012). The relationship between frequency of Facebook use, participation in Facebook activities, and student engagement. *Computers & Education*, *58*, 162-171. doi: 10.1016/j.compedu.2011.08.004.

Junco, R., Heiberger, G., & Loken, E. (2011). The effect of Twitter on college student engagement and grades. *Journal of Computer Assisted Learning*, *27*(2), 119-132. doi: 10.1111/j.1365-2729.2010.00387.x.

Keengwe, J., Kidd, T., & Kyei-Blankson, L. (2009). Faculty and technology: Implications for faculty training and technology leadership. *Journal of Science Education Technology*, *18*, 23-28.

Lemoine, P., & Richardson, M. D. (2013). Cyberlearning: The impact of instruction on higher education. *The Researcher: An Interdisciplinary Journal*, *26*, 57-83.

Lepi, K. (2013). *How social media is being used in education*. Retrieved from http://www.edudemic.com/social-media-in-education/.

Liu, M., Kalk, D., Kinney, L., & Orr, G. (2009). Web 2.0 and its use in higher education: A review of literature. World Conference on E-learning in Corporate, Government, Healthcare, and Higher Education (ELEARN), October 26.

Liu, Z. F., & Chang, Y. F. (2010). Gender differences in usage, satisfaction, and performance of blogging. *British Journal of Education Technology*, *41*, 39−43. doi: 10.1111/j.1467-8535.2009.00939.x.

Long, C. (2009, June 18). *Online social networking for educators: Educators build community and collaboration online.* http://wsww.nea.org/hem/20746.htm.

Mao, J. (2014). Social media for learning: A mixed methods study on high school students' technology affordances and perspectives. *Computers in Human Behavior*, *33*, 213−223. doi: 10.1016/j.chb.2014.01.002.

Martindale, T., & Dowdy, M. (2010). Personal learning environments. In G. Veletsianos (Ed.), *Emerging technologies in distance education* (pp. 177−193). Edmonton, AB: Athabasca University Press.

Mazer, J. P., Murphy, R. E., & Simonds, C. J. (2007). I'll see you on "Facebook." The effects of computer-mediated teacher self-disclosure on student motivation, affective learning, and classroom climate. *Communication Education*, *56*, 1−17. doi: 10.1080/03634520601009710.

McClure, A. (2013). *Social media for retention: Missed opportunities.* Retrieved from http://www.universitybusiness.com/article/social-media-retention-are-colleges-missing-opportunities.

McEwan, B. (2012). Managing boundaries in the Web 2.0 classroom. *New Directions for Teaching & Learning*, *131*, 15−28. doi: 10.1002/tl.20024.

Mishra, T., Lemoine, P., Campbell, K., Mense, E. G., & Richardson, M. D. (2013). Social media and instruction: Irreconcilable differences? In H. H. Yang, Z. Yang, D. Wu, & S. Liu (Eds.), *Transforming K-12 classrooms with digital technology*. Hershey, PA: IGI.

National Science Foundation (2008). *Fostering learning in the networked world: The cyber-learning opportunity and challenge. A 21st century agenda for the National Science Foundation.* Arlington, VA: The National Science Foundation. Retrieved from: http://www.nsf.gov/pubs/2008/nsfD82O4/

nsf08204_1.pdf.

Newmann, F. M., & Wehlage, G. G. (1993). Five standards of authentic instruction. *Educational Leadership*, 50, 8−12.

Nicolini, D., Mengis, J., & Swan, J. (2012). Understanding the role of objects in cross-disciplinary collaboration. *Organization Science*, 23(3), 612−629. doi: 10.1287/orsc.1110.0664.

Pascarella, E. T. (1985). Students' affective development within the college environment. *Journal of Higher Education*, 56, 640−663.

Puzio, E. (2013). Why can't we be friends? How far can the state go in restricting social networking communications between secondary school teachers and their students? *Cardozo Law Review*, 34(3), 1099−1127.

O'Reilly, T., & Battelle, J. (2009). *Web squared: Web 2.0 five years on*. Special report for the Web 2.0 summit. San Francisco CA. Retrieved from http://assets.en.oreilly.com/1/event/28/web2009_websquared-whitepaper.pdf.

Queirolo, J. (2009). Is Facebook as good as face-to-face? *Learning and Leading with Technology*, 57(4), 8−9.

Ramig, R. (2014). *One-to-one computing and learning: Has it lived up to its expectations?* Retrieved from http://www.internetatschools.com/Articles/Editorial/Features/One-to-One-Computing-and-Learning-Has-It-Lived-Up-to-Its-Expectations−95178.aspx.

Roorda, D. L., Koomen, H. M. Y., Spilt, J. L., & Oort, F. J. (2011). The influence of affective teacher-student relationships on students' school engagement and achievement: A meta-analytical approach. *Review of Educational Research*, 81, 493−529. doi: 10.3102/0034654311421793.

Schirmer, J. (2011). Fostering meaning and community in writing course via social media. In C. Wankel (ed.), *Teaching arts and sciences with the new social media: Cutting-edge technologies in higher education* (vol. 3). Bingley, U.K.: Emerald Group.

Schwartz, H. L. (2009). Facebook: The new classroom commons? *Chronicle of Higher Education*, 56(6), B12−13.

Sharples, M., Graber, R., Harrison, C., & Logant, K. (2008). E-safety and Web 2.0 for children aged 11−16. *Journal of Computer Assisted Learning*,

25, 70-84.

Sheldon, P. (2014). *Applying the theory of reasoned action to student-teacher relationships on Facebook*. Paper presented at the annual meeting of the Association for Education in Journalism and Mass Communication (AEJMC), Montreal, Canada.

Shih, C, & Waugh, M. (2011). Web 2.0 Tools for learning in higher education: The presence of blogs, wikis, podcasts, microblogs, Facebook and Ning. In M. Koehler and P. Mishra (Eds.), *Proceedings of Society for Information Technology and Teacher Education International Conference 2011* (pp. 3345-3352). Chesapeake, VA: AACE.

Sim, J. W. S., & Hew, K. F. (2010). The use of weblogs in higher education settings: A review of empirical research. *Educational Research Review*, 5(2), 151-163. doi: 10.1016/j.edurev.2010.01.001.

*Social admissions report* (2013). Retrieved from http://www.theslateonline.com/article/2013/11/su-uses-social-media-to-attract-new-students.

Sturgeon, C. M., & Walker, C. (2009). *Faculty on Facebook: Confirm or deny*. Paper presented at the Annual Instructional Technology Conference, Murfreesboro, TN.

Thompson, C., Gray, K., & Kim, H. (2014). How social are social media technologies (SMTs)? A linguistic analysis of university students' experiences of using SMTs for learning. *Internet & Higher Education, 21*, 31-40. doi: 10.1016/j.iheduc.2013.12.001.

Thompson, D. (2014, June 19). *The most popular social network for young people? Texting*. Retrieved from http://www.theatlantic.com/technology/archive/2014/06/facebook-texting-teens-instagram-snapchat-most-popular-social-network/373043/.

Van Merriënboer, J. J. G., & Stoyanov, S. (2008). Learners in a changing learning landscape: Reflections from an instructional design perspective. In J. Visser & M. Visser-Valfrey (Eds.), *Learners in a changing learning landscape* (pp. 69-90). New York: Springer.

Xie, Y., Ke, F., & Sharma, P. (2008). The effect of peer feedback for blogging on college students' reflective learning processes. *The Internet and Higher Education, 11*(4), 18-25.

Yang, C., & Chang, Y. (2011). Assessing the effects of interactive blogging on student attitudes towards peer interaction, learning motivation, and academic achievements. *Journal of Computer Assisted Learning*, *28*(2), 126−135. doi: 10.1111/j.1365-2729.2011.00423.x.

# 7

# 社交媒体与灾难传播

自然灾难古来有之，然而随着社交媒体的兴起，在全球范围内传播灾难信息则是几秒钟的事。一般人会利用社交媒体发布消息、筹集资金；救援人员会利用社交媒体相互沟通，并与亟需救援的民众保持联络；受灾群众则利用社交媒体联系家人和朋友。对大部分人而言，包括 Facebook、YouTube 和 Twitter 在内的社交媒体，在灾难发生中及发生后为公众及时提供了最新的信息，并能让灾难信息在全球迅速传播。美国联邦应急管理署（Federal Emergency Management Agency）呼吁公众通过发布推文或更新自己的 Facebook 状态来向亲属报平安。社交媒体战略家玛丽·史密斯（Mari Smith）称，现在我们有社交灾难、社交突发事件、社交地震和社交飓风（DiBlasio，2012）。由于任何人都可以通过智能手机在社交网络中发布信息，因此社交媒体对于灾难信息有着病毒式传播的潜力。

社交媒体能够帮助挽救生命。在 2010 年海地地震中，实时"灾难地图"的出现，帮助救援人员定位需要帮助的人。虽然关爱不分国界，但是在一些几乎没有智能手机、互联网费用昂贵或尚未接入互联网的国家，数字鸿沟依旧存在。在海地，大部分人仍然靠收音机或口耳相传的方式来获取新闻，而在那些可以使用互联网的地区,受灾者则会通过社交媒体与亲人联系。

## 灾难的定义

灾难(disaster)指的是"对社区或社会运作造成的严重破坏,它造成广泛的人力、物力、经济或环境损失,且超出受影响地区利用自身资源加以应对的能力"(National Science and Technology Council, 2005, p.21)。根据美国恐怖主义及应对策略全国研究联盟(National Consortium for the Study of Terrorism and Responses to Terrorism, 2012)的定义,灾难传播涉及:(1)政府和应急管理机构通过传统媒体或社交媒体向公众传播灾害信息;(2)记者和受灾群众通过社交媒体或口耳相传的方式发布、分享灾害信息。不同于具有地域性且不可控的灾难,危机是基于组织机构且是人为的(Seeger, Sellnow, & Ulmer, 1998)。危机(crisis)被定义为"扰乱机构或学校的正常运行,威胁机构声誉或机构人员的福祉、财产、财政资源的突发或意外事件"(Zdziarski, 2006, p.5)。例如,洪水和地震属于自然灾害,而枪击则是危机。在本章中,这两个名词将交替使用,因为自然灾害可能会导致危机的出现。

## 灾难传播中的社交媒体

社交媒体正日益成为灾难传播的重要组成部分(Howell & Taylor, 2012)。根据美国红十字会(American Red Cross, 2010)的报告,每六个人中就有一个人使用社交媒体来获取灾难信息。"9·11恐怖袭击事件"表明,当灾难发生时,亟需先进的传播技术来进行灾难传播(Freberg, Saling, Vidoloff, & Eosco, 2013)。

2009年,全美航空公司1549号班机迫降纽约哈德逊河,Twitter是首家发布照片的媒体(Baron & Philbin, 2009)。在2008年印度孟买恐怖袭击事件(Burg-Brown & Mistick, 2012)和2007年美国加州森林火灾事件(Veil, Buehner, & Palenchar, 2011)中,Twitter是现场信息的首要来源之一。2010年海地地震后,尽管许多人无法使用互联网,短信和社交媒体依旧是救援行动中的关键性资源(Burg-Brown & Mistick, 2012)。为了响应社交媒体的号召,智能手机用户每次发送短信"海地"到90999,就会收取10美元的短信费,在十天内为红十字会筹集了2 500万美元。2011年,日本遭遇地震和海啸,人们利用博客抒发情感、互相支持,登入Twitter阅读、传播突发新闻,并通过YouTube观看、分享令人震惊的受灾场景(PEJ New Media Index, 2011)。震后,日本一位Twitter用户向美国驻日本大使约翰·鲁斯(John Roos)求助,他在推文中写道:"千叶市龟田医院需要从30公里(原文如此)外的磐城公立医院转移80名患者。"(Abbasi, Kumar, Filho, & Liu, 2012)

在2011年的飓风艾琳事件中,美国官方首次使用社交媒体来发布关于灾难及应对准备的相关信息(Abbasi et al., 2012)。在2012年的飓风桑迪事件中,110万条推文提到了"飓风"一词,桑迪也因此名列当年Facebook热门话题榜的第二位。即使是在社交网站普及前,人们也会用其他类型的社交媒体获取信息或讨论灾难。在2001年9月11日的恐怖袭击中,博客是人们讨论袭击事件最主要的阵地(Orlando, 2010)。2005年飓风卡特里娜后,人们建立了相关的博客和数据库来搜寻失踪人员的下落。其中一个数据库是寻人启事(PeopleFinder),其页面上有两个按钮:"我在寻找某人"(I'm Looking for Someone)和"我有关于某人的信息"(I Have Information about Someone)。

2007年美国弗吉尼亚理工大学枪击案发生后的20分钟，Facebook页面上出现了一句话："我在弗吉尼亚理工大学还好。"（I'm OK at VT.）仅仅90分钟后，一个能够准确描述这场大屠杀的维基百科页面就诞生了。2007年加州森林火灾期间，人们依靠短信、社交媒体共享信息。海地地震后，由一群自发参与救援的资讯网络专业人士组成的救援小队搭建了OpenStreetMap软件平台，利用受灾地区的谷歌地图及灾后卫星图像，对损坏的建筑物及救援人员所需的其他关键信息进行编码。

根据美国恐怖主义及应对策略全国研究联盟（2012）的标准，公众在灾难中使用社交媒体的理由如下：

- 便利（人人都有一台智能手机）
- 社会规范（亲友都在使用它）
- 来自朋友的推荐
- 顺应潮流（作为一种应对机制）
- 寻求信息
- 及时的信息
- 独有的信息
- 未被过滤的信息
- 确定灾难的量级
- 与家人和朋友保持联系
- 自发组织救援
- 维系一种归属感
- 寻求情感支持和治愈

奥斯汀、费希尔·刘和吉恩（Austin, Fisher Liu, & Jin, 2012）使用社交媒介危机传播（social-mediated crisis communication, SMCC）模型，针对受众如何从社交媒体和传统媒体中获取信息以及在危机中影响其媒介使用的因素展开了研究。结果显示,当危机发生时,受众通常使用社交媒体来获取内幕消息,并与亲朋保持联系;而传统媒体则被用于教育目的。然而,使用社交媒体最主要的动机是获得归属感。2011年日本地震后,人们使用社交媒体来获取积极的心理安慰并与他人保持联系(Howell & Taylor, 2012)。在阅读他人的帖子时,人们能被帖子中流露出的帮助和支持所鼓舞,此时他们不再是受害者或者旁观者。豪厄尔和泰勒(Howell & Taylor)认为,社交媒体是一种参与、监控没有被主流媒体过滤的事件或社会舆论的重要途径。社交媒体具有及时、回应迅速的特点,人们能通过社交媒体向受灾人员提供信息、建立联系、给予帮助、传递爱心,提供一种行之有效的虚拟形式的心理急救(Howell & Taylor, 2012)。

危机发生时,使用社交媒体的另一个原因是没有其他可用的传播渠道。曹和朴(Cho & Park, 2013)研究了2011年日本地震期间的社交媒体使用后发现,地震发生后,由于固定电话和手机均无法接通,人们只能依靠社交媒体传递信息。事实表明,Twitter上地震消息的发布远比传统媒体快。因此,曹和朴(2013)认为,在危机发生期间,相比官方消息源,社交媒体用户更依赖点对点通信和信息导向型的网站。他们还发现,危机期间,Twitter用户发布的帖子鲜少与自身相关,而是更多地提供有关危机的信息。在灾难发生时,Twitter上的回复量下降,但转发量会上升(Cho & Park, 2013)。

在自然灾害中,人们利用社交媒体来搜寻或报告失踪人员,请求支援,帮助重建受灾地区。例如,2010年海地地震中创建的谷歌寻人项目(Google Person Finder),目的是为了帮助人们在自然灾害和人道主义灾难后与朋友和亲人重新取得联系。该网站在2010年智利地震、2011年3·11日本大地震及波士顿马拉松爆炸事件中都被使用。事实表明,在2013年美国波士顿马拉松爆炸事件中,社交媒体既是应急工具,又是抒发悲伤情感的平台(Bellantoni,2013)。在报道事件时,Twitter的传播速度快于广播和电视,并且既有真实的照片又有目击者的证词,因此既报道了事件,又增强了社会凝聚力并表达了同情。

约尔特和金(Hjorth & Kim,2011)研究了移动社交媒体如何在日本地震中保持公众之间的及时联系。地震后,手机与电话线路中断,而社交媒体则通过图像、文本信息和口头交流,提高了公民参与度并拓展了媒体报道的形式。社交媒体"产生了一种新的集体情感力量,让我们感觉到联系更加紧密"(p.553),并且"可以超越语言、地理和时间的边界,让每个人都能参与其中"(p.554)。虽然新媒体并没有引发全新的变革,但约尔特和金(2011)认为,它们改变了我们定义和体验事件的方式。这种体验和媒介形式有点类似。比如,短信和Twitter公告(post)类似于明信片的功能,"公告"意味着传递。事实上,在1923年的东京地震中,最多被使用的媒介是明信片。图片式的明信片意在利用真实的视觉图像来向公众告知这个城市发生了什么。明信片就像社交媒体一样,可以在不同的时空中,经过多人之手传递。唯一的区别是,明信片不是即时的。灾难发生时,人们会使用社交媒体作为一种新的咨询平台,但正如约尔特和金(2011)强调的,人们对传统媒介形式的需求尤其显著,包括面对面交流。

拉克伦、斯彭斯和林(Lachlan,Spence,& Lin,2014)聚焦桑迪飓风事件中的社交媒体网站Twitter,对事件发生之前的推文进行了内容分析。他们认为,在危机期间,Twitter是重要的新闻和信息来源,例如提供了与受灾者需求相关的信息(Lachlan,Spence,& Lin,2014)。和其他社交网站一样,Twitter通常是人们遭遇胁迫和不确定事件时与他人保持联系的工具。在灾难传播中使用这类媒介,可能有助于缓和受灾者、受灾者亲属和其他关心灾难的人们的消极情绪。此外,瓦斯科(Wasike,2013)指出,Twitter也可能会使用传统媒体常用的框架技巧,包括强调新闻中的人情味、冲突和经济利益等元素,而其他社交网站则很少使用这类方法。

## 渠道互补理论和媒介依赖理论

根据渠道互补理论(channel complementarity theory;Dutta-Bergman,2006),受众会根据他们所需要的功能来选择特定的媒介。这些媒介往往与受众的观念和思维方式相一致,从而增强了受众的信念。瑞安(Ryan,2013)认为,灾难的类别决定了人们寻求信息的方式。灾难发生时,不同的媒介发挥着不同的功能。尤里克和西尔维斯特(Juric & Sylvester,2007)以美国学生及留学生为研究对象,采取调查和焦点小组访谈的方法,对2005年飓风卡特里娜发生时的媒介使用差异进行探析。结果表明,对大部分人而言,电视(图像、大尺度的镜头;人们想看的关于灾害的照片)是首选,其次是互联网(能够指定关键词,避免和电视重复的报道)。只有当灾难使以上的方式断电时,他们才会使用电池供电的收音机获取信息。尤里克和西尔维斯特(2007)认为,灾难期间,媒介与受众的关系会变得更加紧密。媒介不仅为受众提供所需的信息,还提供了情感支持和归属感。根据媒介依赖理论(media dependency theory;BallRokeach & DeFleur,1976),人们会对那些能够满足自身需求的媒介产生依赖。使用媒介的时间越长,个人对媒介内容的依赖性就越强。如果媒介内容能够满足更多受众的需求,那么受众对媒介的依赖会更强。灾难期间,人们对某一媒介越熟悉,对它的依赖程度就越高。

社交媒体具有可参与性、开放性、互动性、社区性和连接性的特质。因此,无论灾难发生时还是灾后,社交媒体都是流行的选择(Mayfield,2006)。就算记者不在场,关于危机的新闻也能够通过社交媒体共享,并传达给数以百万计的人。然而,拥抱社交媒体并不意味着停止使用主流媒体。这两类消息源可以相辅相成(Wright & Hinson,2009)。在 2011 年澳大利亚洪灾和日本地震期间,人们都是先观看电视新闻,再转向 Facebook 作为第二个关键信息源(Howell & Taylor,2012)。在应对危机时,从业者不应忽视传统媒体和口耳相传(的传播方式)。奥斯汀、费希尔·刘和吉恩(2012)的研究显示,人们最早是通过口头传播得知危机,其次是电视,最后才是 Facebook。可见,传统媒体仍然具有很高的可信度。

通过测量社交媒体的使用行为,可以观测灾害发生时个人是如何使用社交媒体进行互动的。根据社交媒介危机传播模型(Jin & Liu,2010),危机发生前、发生时和发生后,有三类人在生产和消费信息:(1)社交媒体上有影响力的用户,他们将发布的第一手危机信息提供给其他用户;(2)社交媒体上的关注者,他们消费这些信息;(3)社交媒体上不活跃的用户,他们通过口耳相传的方式从关注者或者信息创建者那里了解信息。灾难发生时,活跃在社交媒体上的用户大多是那些关注者,即第二类人(Austin,Fisher Liu,& Jin,2012)。

# 建构危机信息

一些研究(例如,Schultz,Utz,& Goritz,2011;Cho & Gower,

2006)发现,灾难发生时,媒介本身要比信息更重要。公众无法了解危机事件的客观事实,只能通过媒体或新闻所构造的内容对其进行感知。换言之,媒体报道危机的方式会影响公众对危机的感知。例如,人情味框架(human interest frame)会让人们认为危机是严重的、紧急的、危险的。记者们常说:"只要有流血,就能上头条。"(If it bleeds, it leads.)在曹和高尔(Cho & Gower, 2006)的研究中,研究对象阅读新闻内容时,阅读到人情味新闻的人会比未阅读到的人表现出更明显、更强烈的情绪反应。曹和高尔认为,过于戏剧化的报道和情绪化的新闻会刺激人的心理脉冲,并且会使人们在感知某个危机事件时,对涉事的组织机构产生负面的认知。

韦斯特曼、斯彭斯和拉克伦(Westerman, Spence, & Lachlan, 2009)认为,对威胁性新闻的迷恋是一种自我保护的手段。通常,对灾难的新闻故事会引起人们的关注,而新闻故事中不断增加的恐惧感则进一步增加了公众对报道的关注度(Young, 2003)。根据例证理论(exemplification theory; Zillmann, 2002),比起抽象的、象征性的、与情感无关的例证,那些具体的、形象的、能激发情感的例证更能影响人们对事件的感知。例证促使人们采取保护措施来应对这种不断增加的威胁。那些亲身经历过灾难的人也会对社交媒体上报道的新闻更感兴趣。以往的研究(Choi & Lin, 2009b; MacInnis, Rao, & Weiss, 2002; Claeys & Cauberghe, 2013)表明,危机参与程度高的人会比参与程度低的人更深入地审视危机信息。

## 危机中的媒介可信度

灾难期间,人们会借助社交媒体来寻求即时、准确的信息

(Bates & Callison，2008)。2008年中国汶川地震(在美国)的首条报道来自Twitter(Mills，Chen，Lee，& Rao，2009)。对2007年美国加州森林火灾中受灾居民的调查显示,很多人认为主流媒体并没有及时提供足够的信息。因此,人们会转向社交媒体寻求信息(Mills et al.，2009)。

拉克伦、斯彭斯、爱德华兹、雷诺和爱德华兹(Lachlan，Spence，Edwards，Reno，& Edwards，2014)根据更新速度,考察了社交媒体信息可感知的信任度。研究者认为,如果有人重复地从特定信源或媒介获取信息,那么他们对其的信任度会越来越高。由于社交媒体网站具有持续且及时更新信息的特质,就自然而然地增加了人们的信息寻求行为。基于上述研究者的假设,这能够增强信任和可感知的可信度。因此,人们对灾难或危机的风险感知程度应该基于他们对信源和/或媒介的信任度。这对社交媒体信息的重要度和可信度的判断十分关键,因为它们将决定信息接收方的行为。如果信息是某种警告,那么通过社交媒体感知到的灾难严重性和社交媒体信息的可信度将会影响信息接收方即将采取的安全预防措施以及对是否要发布紧急通知、如何发布和向谁通知的判断。这对高校来说尤为重要,因为他们已经习惯了大范围地使用电话、短信提醒、电子邮件、更新社交媒体这些应急机制。然而,鲜少有研究探讨技术选择是如何影响人们对危机事件严重性的感知。谢尔顿(Sheldon，2015)研究了在校园危机事件(大规模枪击事件 vs.龙卷风)中,用来提醒学生所采用的技术(短信 vs.社交媒介)是如何影响学生对事件严重性的感知、把信息分享给亲朋好友的意图以及会选择何种渠道来分享。一项以177名大学生为考察对象的研究显示,人们认为学校通过短信发出的危机警报比通过Facebook发出的更严重。而无论面对何种类型的危机,无论用何种媒介发送警报，学生们表示他们首先会告知的是在身边的人。与其他媒介依赖的研究结果一致，我们越熟悉某种媒介，在灾难发生时就会越依赖它。

## 社交媒体的缺点

根据美国恐怖主义及应对策略全国研究联盟（2012）的报告，在灾难发生时，造成公众可能不使用社交媒体的原因如下：
- 隐私和安全担忧
- 准确性担忧
- 访问障碍（停电；数字鸿沟）
- 知识匮乏

大规模灾难期间，事件最初是由目击者通过手机发布，紧接着相关信息在社交媒体上传播，随后再是主流媒体的报道（Oh, Agrawal, & Rao, 2011）。奥等人（Oh et al.）认为，当地人有可能会成为首批救援人员，因为他们对当地情况非常熟悉，这恰恰是来自受灾地区之外的专业救援人员所缺乏的。然而，这也有可能会导致谣言、误报、传闻的传播。大规模的社会危机往往会导致信息过载，其中就包括那些不准确的内容。奥、阿格拉沃尔和拉奥（Oh, Agrawal, & Rao, 2013）认为，应对危机的重要任务之一就是控制谣言，并且通过各种传播渠道尽可能快速地向受灾地区传达当地的、可靠的信息。

## 公关从业人员指南

越来越多的公关从业人员开始使用社交媒体平台来衡量公众

情绪、预防危机或应对危机。豪厄尔和泰勒（Howell & Taylor）认为，一个缺乏全面数字化战略的计划不是一个完整的危机应对计划。危机管理者需要制定新传播工具的使用指南（Howell & Taylor, 2012）。此外，主流媒体也会将 Facebook 作为新闻报道的消息源。为了赢得公民的信任，灾难发生前媒体就应建立相应的官方页面（Howell & Taylor, 2012）。据美国联邦应急管理署（FEMA, 2012）的《自然灾害应对及恢复中的社交媒体应用指南》(*Social Media for Natural Disaster Response and Recovery*)显示，在灾难管理中，社交媒体通常肩负着"为公众应对紧急情况做准备、监测需要帮助的人、提醒和警告公众、支持救灾和重建工作、收集数据、及时向公众更新新闻和信息。"的职责。社交媒体应该应用于灾难的预防/减轻、应对和重建阶段。红十字会的一项调查（American Red Cross, 2010）显示，由于这些组织的职责之一是监测、应对紧急请求，因此公众希望能看到救灾组织利用社交媒体发布的消息。相关数据表明，有 74% 的人希望能在一小时内获得帮助（American Red Cross, 2010）。联邦应急管理署（2012）的指南督促救灾组织遵循"一个声音，多种渠道"的原则，来保证发布在社交媒体上的消息"简洁、切题"。然而，根据目标任务，不同的平台可能比其他形式更有优势。Twitter 信息发布更迅速，更适合发布、更新短新闻；而 Facebook 的结构鼓励建立更多社群和讨论（FEMA, 2012）。尽管一些组织已经建立了常规的 Facebook 页面，联邦应急管理署还是鼓励每个人都建立一个 Facebook 页面。与私人档案不同，Facebook 页面是公开的，人们不需要与这个账号成为朋友就能给它"点赞"。

美国亚利桑那州立大学（Arizona State University）的研究者（Abbasi et al., 2012）开发了一套名为"TweetTracker"的 Twitter 监

测与分析系统,可以便捷地跟踪并获取灾害信息,以协助救援人员做出有效决策。阿巴西等人(Abbasi et al.,2012)通过真人角色扮演的实践测试了这套系统。在测试中,受害者利用社交媒体来寻求帮助,这些请求经由救助系统(TweetTracker)的处理生成了一份报告以供救援人员参考。基于这项测试,阿巴西和他的同事们总结了以下经验。首先,必须采用正确的方法从成千上万条推文中筛选出相关的推文。其次,人们需要在推文中分享他们的位置(地理位置)。但出于隐私担忧和缺乏对位置功能的了解,只有不到5%的用户在他们的推文中提供了地理位置信息(Abbasi et al.,2012)。

奥兰多(Orlando,2010)认为社交媒体并不只是向公众输出信息的一个渠道,它们还会从公众那里获取信息和资源。奥兰多还呼吁商界使用社交媒体,基于多来源及大量数据的"协同知识"要比单一来源的知识更准确。无论有没有官方消息,人们总是会在社交媒体上谈论一些事件(Orlando,2010)。

佩奇、弗雷伯格和扎林(Page, Freberg, & Saling, 2013)则呼吁包括政府机构在内的危机管理者,不仅在救灾和重建阶段,更应在危机预备阶段就开始使用社交媒体。相关准备工作包括制定危机计划、使用各种形式的媒介来改进传播体系。佩奇等人表示,在危机期间,救灾人员和官员如何提供及时准确的信息仍然是一个有待研究的领域。而预先制定危机计划有助于减少公众的恐惧和不确定性,同时可以提高救灾人员的可信度。

弗雷伯格、扎林、维多洛夫和厄斯科(Freberg, Saling, Vidoloff, & Eosco, 2013)构建了新兴媒体危机价值模型(Emerging Media Crisis Value Model, EMCV),用以解释危机信息的一般功能和次级功能。有价值的信息具有传播迅速、可信、准确、简单、完整及广泛传播的特点。然而,"可信的"消息并不适用于所有自然灾害,比如飓风。该模型指出了在危机中使用社交媒体的三种最佳实践方式:(1)整合更新多媒体信息和链接;(2)正确使用标签和关键字;(3)注意官方信息和对话式信息更新之间的平衡。

佩奇等人(2013)使用 EMCV 模型界定了什么是一条"好"的危机信息。他们分析了两个案例:2011 年席卷美国东海岸的飓风艾琳和 2012 年在奥罗拉的科罗拉多剧院的枪击案(一个名叫詹姆斯·霍姆斯[James Holmes]的人在午夜持枪进入剧院并造成 12 人死亡)。佩奇等人(2013)发现,当对自然灾害和人为灾难中的传播进行比较时,除了责任归属外,其余的信息价值在衡量上并没有差别。在人为的危机中,公众会寻找一个承担责任的政党,因为危机本来是可以避免的。根据调查结果,佩奇等人(2013)基于平台的受欢迎度,建议人们使用 Instagram、Pinterest 和 Tumblr 这类视觉的社交平台,因为这些平台比文本平台更流行。例如,尽管美国目前并未发生严重的危机,联邦应急管理署仍会在社交媒体上保持活跃。在它的 Facebook 主页上,联邦应急管理署会定期发布一些如何应对灾难的信息,通过视频和照片来告知人们未雨绸缪的重要性。综上所述,在灾难发生的各个阶段,社交媒体发挥着重要的作用,并且不同的社交媒体工具能被同时使用(FEMA,2012)。

# 参 考 文 献

Abassi, M. A., Kumar, S., Filho, J. A. A, & Liu, H. (2012). *Lessons learned in using social media for disaster relief—ASU crisis response game*. In SBP's 12 Proceedings of the 5th International Conference on Social Computing, Behavioral-Cultural Modeling and Prediction, pp. 282-289.

American Red Cross (2010). *Social media in disasters and emergencies*. American Red Cross, Washington, DC.

Austin, L., Fisher Liu, B., & Jin, Y. (2012). How audiences seek out crisis information: Exploring the social-mediated crisis communication model. *Journal of Applied Communication Research*, 40, 188–207. doi: 10.1080/00909882.2012.654498.

Ball-Rokeach, S. J., & DeFleur, M. L. (1976). A dependency model of mass media effects. *Communication Research*, 3, 3–21. doi: 10.1177/009365027600300101.

Baron, G. & Philbin, J. (2009). Social media in crisis communication: Start with a drill. *Public Relations Tactics*, 16(4), 12.

Bates, L., & Callison, C. (2008). *Effect of company affiliation on credibility in the blogosphere*. Paper presented at the Association for Education in Journalism and Mass Communication Conference, Chicago, IL.

Bellantoni, C. (2013). *In face of disaster, social media helped spread news and connect Bostonians*. Retrieved from http://www.pbs.org/newshour/bb/media/jan-june13/dd_04-16.html.

Burg-Brown, S., & Mistick, D. (2012). One question, two members. *Journal of Property Management*, 77(5), 9.

Cho, S. E., & Park, H. W. (2013). Social media use during Japan's 2011 earthquake: How Twitter transforms the locus of crisis communication. *Media International Australia*, 149, 28–40.

Cho, S., & Gower, K. K. (2006). Framing effect on the public's response to crisis: Human interest frame and crisis type influencing responsibility and blame. *Public Relations Review*, 32(4), 420–422. doi: 10.1016/j.pubrev.2006.09.011.

Choi, Y., & Lin, Y-H. (2009a). Consumer responses to Mattel product recalls posted on online bulletin boards: Exploring two types of emotion. *Journal of Public Relations Research*, 21(2), 198–207. doi: 10.1080/10627260802557506.

Choi, Y., & Lin, Y-H. (2009b). Consumer response to crisis: Exploring the concept of involvement in Mattel product recalls. *Public Relations Review*, 35(1), 18–22. doi: 10.1016/j.pubrev.2008.09.009.

Claeys, A. -S., & Cauberghe, V. (2013). What makes crisis response strategies work? The impact of crisis involvement and message framing. *Journal of Business Research*, *67*, 182-189. doi: 10.1016/j.jbusres.2012.10.005.

DiBlasio, N. (2012, August 30). Relief groups try tweets, apps to spread the news. *USA Today*, p 4A.

Dutta-Bergman, M. J. (2006). Community participation and Internet use after September 11: Complementarity in channel consumption. *Journal of Computer-Mediated Communication*, *11*, 469-484. doi: 10.1111/j.1083-6101.2006.00022.x.

FEMA (2012). Social media for natural disaster response and recovery. Retrieved from http://www.utc.edu/safety-risk-management/pdfs/website/conference/social-media-handouts.pdf.

Freberg, K., Saling, K., Vidoloff, K. G., & Eosco, G. (2013). Using value modeling to evaluate social media messages: The case of Hurricane Irene. *Public Relations Review*, *39*(3), 185-192. doi: 10.1016/j.pubrev.2013.02.010.

Hjorth, L., & Kim, K. (2011). The mourning after: A case study of social media in the 3.11 earthquake disaster in Japan. *Television & New Media*, *12*(6), 552-559. doi: 10.1177/1527476411418351.

Howell, G. V. J., & Taylor, M. (2012). *When a crisis happens, who turns to Facebook and why?* Retrieved from http://www.deakin.edu.au/arts-ed/apprj/articles/12-howell-taylor.pdf.

Jin, Y., & Liu, B. F. (2010). The blog-mediated crisis communication model: Recommendations for responding to influential external blogs. *Journal of Public Relations Research*, *22*, 429-455. doi: 10.1080/10627261003801420.

Juric, P., & Sylvester, J. (2007). Mass media use during a natural disaster: Louisiana State University students and Hurricane Katrina. *Southwestern Mass Communication Journal*, *22*, 85-96.

Lachlan, K. A., Spence, P. R., Edwards, C., Reno, K., & Edwards, A. (2014). If you are quick enough, I'll think about it: Information speed and trust in public health organizations. *Computers in Human Behavior*, *33*(2), 377-380.

Lachlan, K. A., Spence, P. R., & Lin, X. (2014). Expressions of risk awareness and concern through Twitter: On the utility of using the medium as an indication of audience needs. *Computers in Human Behavior*, *35*, 554–559. doi: 10.1016/j.chb.2014.02.029.

MacInnis, D. J., Rao, A., & Weiss, A. (2002). Assessing when increased media weight of real-world advertisements helps sales. *Journal of Marketing Research*, *39*(4), 391–407. doi: http://dx.doi.org/10.1509/jmkr.39.4.391.19118.

Mayfield, A. (2006). *What is social media?* Spannerworks. Retrieved from http://www.spannerworks.com/fileadmin/uploads/eBooks/What_is_Social_Media.pdf.

Mills, A., Chen, R., Lee, J., & Rao, H. R. (2009). *Web 2.0 emergency applications: How useful can Twitter be for an emergency response?* Retrieved from http://denmanmills.net/web_documents/jips_mills.etal._2009.07.22_finalsubmission.pdf.

National Consortium for the Study of Terrorism and Responses to Terrorism (2012). *Social media use during disasters: A review of the knowledge base and gaps.* Retrieved from http://www.start.umd.edu/sites/default/files/files/publications/START_SocialMediaUseduringDisasters_LitReview.pdf.

National Science and Technology Council (2005, June). *Grand challenge for disaster reduction: A report of the Subcommittee on Disaster Reduction.* Washington, D.C.: National Science and Technology Council, Executive Office of the President, Washington, D.C. Retrieved from http://www.nehrp.gov/pdf/grandchallenges.pdf.

Oh, O., Agrawal, M., & Rao, H. R. (2011). Information control and terrorism: Tracking the Mumbai terrorist attack through Twitter. *Information Systems Frontier*, *13*, 33–44.

Oh, O., Agrawal, M., & Rao, H. R. (2013). Community intelligence and social media services: A rumor theoretic analysis of tweets during social crises. *MIS Quarterly*, *37*, 407–426.

Orlando, J. (2010). Harnessing the power of social media in disaster response. *Continuity Insights*. Retrieved from http://www.continuityinsights.com/articles/2010/09/harnessing-power-social-media-disaster-response.

Page, S., Freberg, K., & Saling, K. (2013). Emerging media crisis value model: A comparison of relevant, timely message strategies for emergency events. *Journal of Strategic Security*, *6*, 20-31. doi: 10.5038/1944-0472. 6.2.2.

PEJ New Media Index (2011, March 14-18). In social media it's all about Japan. *Pew Research Center's Project for Excellence in Journalism*. Retrieved from http://www.journalism.org/2011/03/24/social-media-its-all-about-japan/.

Ryan, B. (2013). Information seeking in a flood. *Disaster Prevention and Management*, 22, 229-242.

Schultz, F., Utz, S., & Goritz, A. (2011). Is the medium the message? Perceptions of and reactions to crisis communication via Twitter, blogs and traditional media. *Public Relations Review*, *37*(1), 20-27. doi: 10.1016/j.pubrev.2010.12.001.

Seeger, M. W., Sellnow, T. L., & Ulmer, R. R. (1998). Communication, organization, and crisis. *Communication Yearbook*, *21*, 231-275.

Sheldon, P. (2015). *Alerting students about a crisis: Technology preferences and secondary crisis communication.* Presented at the International Communication Associationconference. San Juan, Puerto Rico.

Veil, S. R., Buehner, T., & Palenchar, M. J. (2011). A work-in-process literature review: Incorporating social media in risk and crisis communication. *Journal of Contingencies and Crisis Management*, *19*, 110-122. doi: 10.1111/j.1468-5973.2011.00639.x.

Wasike, B. S. (2013). Framing News in 140 Characters: How Social Media Editors Frame the News and Interact with Audiences via Twitter. *Global Media Journal: Canadian Edition*, *6*(1), 5-23.

Westerman, D., Spence, P. R., & Lachlan, K. A. (2009). Telepresence and the exemplification effects of disaster news. *Communication Studies*, *60*, 542-557. doi: 10.1080/10510970903260376.

Wright, D. K., & Hinson, M. D. (2009). *An analysis of the increasing impact of social and other new media on public relations practice*. International Public Relations Research Conference, Miami, FL. Retrieved from www.instituteforpr.org/wp-content/up-loads/Wright_Hinson_PR_Miami.pdf.

Young, J. R. (2003). The role of fear in agenda setting by television news.

*American Behavioral Scientist*, 46, 1673 – 1695. doi: 10. 1177 / 0002764203254622.

Zdziarski, E. L. (2006). Crisis in the context of higher education. In K. S. Harper, B. G. Paterson, & E. L. Zdziarski (Eds.), *Crisis management: Responding from the heart.* Washington, D. C.: National Association of Student Personnel Administrators.

Zillmann, D. (2002). Exemplification theory of media influence. In J. Bryant & D. Zillmann (Eds.), *Media effects: Advances in theory and research* (pp. 213–245). Mahwah, NJ: LEA.

# 8

# 社交媒体与广告

社会化营销的概念早已有之。但在过去的十年间,互联网一直发挥着辅助传统媒体进行产品推广和服务的作用。技术改变了个人的购买决策、业务开展和与人互动的方式。社交网站拉进了企业与公众之间的距离,并产生了一种新方式来传播公司的品牌形象。本章探讨了社交媒体网站广告的利弊,并基于此进一步为如何在 Facebook、Twitter、YouTube、Pinterest 和 LinkedIn 等平台发布广告提供指导。

## 社交媒体上的广告

与昂贵的标准线上广告相比,企业与消费者之间的沟通方式,正因为社交媒体平台的发展而产生变革(Megna,2009),而这种趋势往往伴随着在线社区的发展而生。在线社区能为公司的品牌宣传提供更大的受众群体。从 Facebook 上的推荐帖和用户定制广告,到在 Twitter 上发布热门品牌话题和广告推文,公司有很多种方式来接触到现有客户和潜在客户。在社交媒体被广泛使用的当下,广告商需要在这些网站上向消费者展开营销。

虽然社交媒体广告通常与传统媒体广告相伴,但社交媒体的交互性、定制化和社交互动这三大特色已经改变了广告业务(Hill & Moran, 2011)。交互性被定义为"通信技术创造中介环境的程度,在该环境中,参与者能够同步或异步地进行相互沟通(包括一对一、一对多和多对多),并参与互惠信息交换(三级依赖)"(Kiousis, 2002, p.372)。交互性能够产生更高的参与度(Bucy, 2003)和来源可信度(Fogg, 2003)。品牌或公司的 Facebook 页面就是一个展示企业是如何轻松地与客户进行互动的例子。埃马努埃利(Emanuelli, 2012)提出了几个建议以帮助增加 Facebook 的页面互动:与粉丝进行互动,分享有趣的内容,分享相关新闻和文章,使用日常语言。例如,耐克(Nike)的 Facebook 页面使用了标志(logo)作为头像,并且将耐克的口号"Just do it"设为封面,这使人们能够更容易地辨识出品牌。鉴于人们乐于看到公司最关心的是人本身,埃马努埃利(2012)还鼓励企业发布个人图片和视频。

另一个增加用户信心的方法是在 Facebook、Twitter 或 LinkedIn 上分享公司最新的业务成果。每个人都希望能看到他们喜爱的品牌获取成功。除了页面的设置,企业还可以投放传统的在线展示广告。目前,Facebook 是最受欢迎的广告投放目标网站。虽然大多数人坦言,他们从不点击 Facebook 上的广告,但是这些广告仍然有一定的效果。2012 年,Facebook 请 Datalogix 公司衡量一亿个家庭的消费习惯,结果表明他们确实购买了在 Facebook 上看到的广告产品。在某些情况下,企业的收益能达到广告费用支出的三倍之多(Manjoo, 2013)。

社交媒体的另一个特点是定制化,它能根据互动中的反馈采取不同方式应对客户和潜在买家(Peppers, Rogers, & Dorf, 1999)。这有助于提高信息的可信度,并通过定位于特定的细分人群而非普罗大众来降低运营成本。在 Twitter 上,"标签"的设置能够便于人们根据各自的兴趣进行搜索（Hill & Moran，2011）。Twitter是向小众市场开展营销的一个不错的选择，而Facebook面向的则是广泛的公众。

最后，社交媒体能让用户进行社交互动以及与好友、粉丝的持续交流。用户不仅会被他们所重视的人影响，更重要的是，他们也可能被网络中的好友或粉丝所影响。这是社交媒体广告与传统广告的一个很大的区别。社交媒体能够让营销人员与社交媒体用户进行双向对话，也可以让用户互相交谈、相互影响。例如在一些在线点评服务中，任何人都可以查看或推荐产品及服务。

## 优点

社交媒体广告具有成本低、速度快、范围广和双向沟通的优势。首先，在社交媒体上发布广告的费用不贵。大多数的社交网站都是可以免费访问的。发布广告的费用相较于其他媒介形式要低得多。这意味着企业可以在少量甚至没有现金投入的前提下接触到目标受众。无论目标受众在家还是在遥远的地方，企业都能接触到他们，也能向他们提供比传统形式广告更多的信息。

此外，在社交媒体上发布广告既快速又简单。根据近期的一份报告（Stelzner，2014），84%的市场营销人员每周仅需花费六个小时来更新其社交媒体账户。一些推荐的改进策略包括引用名言警句、使用表情符号、发布提问帖、提供赠品和举办比赛、用语幽默和讲笑话、附带其他优质内容的链接、使用视觉图像和对工作人员进行访谈等。

社交媒体的病毒式传播使得有趣的广告能够在较大范围的人际网络中迅速传播（Weinberg，2009）。病毒营销（viral advertising），或蜂鸣营销（marketing buzz），是一种在社交媒体上推广品牌的非常流行和简单的方法。用户经常分享有趣或独特的视频片段,例如百威(Budweiser)制作的超级碗广告Puppy Love。此外,社交媒体广告还能让企业通过接触对品牌提供的产品与服务最感兴趣的人,来精准地定位特定市场。而消费者则可以在社交平台上投诉,这有助于企业了解并回应消费者的不满(Gommans, Krishnan & Scheffold, 2001)。因此,企业就有了即时处理评论、公开道歉、立即采取更正措施的优势。相应的,企业也可以使用社交媒体来研究消费者的兴趣和行为模式。

**缺点**

社交媒体广告并不是完美的,它的缺点包括信任和隐私问题、商标和版权问题以及消费者即刻留下负面反馈的能力(Weinberg, 2009)。此外,更新社交媒体上的相关内容也可能比较耗时。并不是每个公司都有足够的资源专门雇用一个员工来管理所有的帖子、回复评论、回答关于产品的问题,同时深入地了解消费者。此外,消费者可以留下负面的反馈。相关研究发现,社交媒体上的负面帖子造成的影响比正面帖子高出五倍(Corstjens & Umblijs, 2012)。由于大多数互联网公司会从用户那里收集大量的个人数据并将其出售给广告商,因此消费者可能会抵制个性化广告,并将其视为可怕或令人讨厌的事物(Tucker, 2014)。

# 如何在社交媒体上做广告

首先,社交媒体广告应是有意义的、有用的,还要是友善的、有吸引力的,以便于读者向他们的朋友分享。参与是社交媒体广告的关键。例如,如今许多餐厅通过社交媒体来获取客户的反馈,并据此作出改进；一些企业会发布他们提供的食品的照片以吸引食客；一些夜总会在社交媒体上宣传周末和白天特价，来获取更多的流量。因为消费者总是在社交媒体上保持在线（Powers, Advincula, Austin & Graiko, 2013），所以通过社交网站来直接向消费者传达信息是一种非常有效的方式。

其次，企业可以组织一些有不同奖金和奖励的比赛或游戏。例如，可以让用户在时间轴（timeline）上发布照片，得到最多"赞"的人能够获得奖品。企业必须要记住的一件事情是，他们必须要经常在社交媒体上发布内容，这样才不会被人们忘记。类似于传统广告，在社交媒体上发布的广告文案需要简短、有趣，目的是号召人们采取行动。同时也可以提供一些奖励或促销，例如，Facebook 上的 Hello Fresh 广告不仅给首次购买的消费者提供 30 美元的优惠，同时还附带 HelloFresh.com 的网站链接，人们可以点击链接在线订购食物。

社交媒体还是品牌社群的容身之所（Habibi, Laroche, & Richard, 2014）。品牌社群会潜在地影响消费者对品牌的信任度，因此不容小觑。品牌社群的定义强调，这是一个"专业的、无区域界限的，并且基于品牌爱好者的社交关系而建立的社群"（Muniz & O'Guinn, 2001, p.412）。这些社交关系包括四种关系：消费者与产品、消费者与品牌、消费者与公司以及消费者与其他消费者（Habibi et al., 2014）。所谓的消费者与其他消费者的关系，即例如在 Facebook 上，即使你不是某个品牌的粉丝，只要你有好友关注了该品牌，那么它也同样可以定位到你。此外，点评服务也是品牌社群的其中一个部分。消费者能够通过这些品牌社群相互交流来获取更多的信息，要求并期待从品牌中获得更多（Habibi et al., 2014）。

## 在Facebook上做广告

Facebook在线上社交网络中占据主导地位,对所有大中小企业而言都已经成为一个不可或缺的在线营销工具。根据纳拉亚南等人(Narayanan et al.,2012)的研究,Facebook为企业提供了三大便利之处:一个能让现有或潜在消费者通过"点赞"成为其"粉丝"的品牌主页;能在特定粉丝群体内推送特定帖子的社交功能,而且这些帖子同时能够被品牌粉丝的"朋友"看到;此外,如前所述,Facebook还允许企业通过访问用户的人口统计信息来更精准地定位特定受众。公司通过利用粉丝页面、推送帖子和发布广告,公司能够有针对性地开展如打折、产品或服务优惠券及线下促销等活动。通常,这些优惠活动只提供给这一页面的"粉丝",这促使潜在客户去"点赞"成为粉丝以获取特别优惠。

## 在Twitter上做广告

在谈及与其他用户之间的联系时,Twitter与Facebook有一些不同之处。在Twitter上任何用户都可以发布由140个字符所构成的推文,并将其展示给关注他的粉丝们。Twitter的"转发"功能允许所有的用户——无论是否关注帖子的原始发布者——都可以"转发"原始推文,系统还能将信息发布到其"关注"列表,即时间轴。使用Twitter进行营销不仅对及时发布关键信息起到至关重要的作用,同时也使公司能够接触到当前与潜在的新客户。虽然Twitter的粉丝基数较小,但它能够让企业接触到广泛的受众,且速度通常快于Facebook。

目前,在Twitter上做广告包括以下几个步骤。首先,企业可以选择一个活动目标,如通过收藏、转发、回复及下载企业移动应用程序获得更多的用户关注。其次,广告商可以基于地理位置、设备、性别、语言、关键词和用户的兴趣取向,来选择他们的目标受众群体。Twitter 还有一个 Analytics(分析)服务,可以帮助企业更多地了解他们的粉丝。根据网站规定,Twitter 上禁止发布与酒精饮料、金融服务、赌博、健康与制药产品服务及政治运动相关的广告。在广告产品方面,企业可以有几种选择。他们可以选择一个推广账户,以关注成本(Cost-Per-Follow,CPF)为基础定价,这样做的目标是增加在 Twitter 上的粉丝数量。企业也可以选择使用推文推广(Promoted Tweet)选项,所支付的广告推文能够显示在不是其粉丝的用户时间轴上。广告通过拍卖出售,意味着被推广的推文是出价最高的推文。在任何时间,只会有一条推广的推文会在用户的时间轴上出现,并且被明确标识为推广。第三种选择是热门推广(Promoted Trends)——热门主题会出现在首页左侧,并被标记上推广的提示图标。根据 Twitter 的介绍,推荐热点有助于促进品牌与消费者之间的对话。

不少研究已经证实了 Twitter 广告的有效性。2012 年,维特柯帕、钟·胡恩和瓦尔德布格(Witkemper,Choong Hoon,& Waldburger)研究了社交媒体是如何改变消费者和球迷与运动员保持联系的方式。通过分析人们与运动员保持联系的动机,并考虑到粉丝的经济实力和能接近运动员的程度等方面的限制条件,维特柯帕等人得出了为什么某些运动员在 Twitter 上能比其他运动员更具有市场价值的研究结论。他们发现,一个拥有越多粉丝的 Twitter 用户,越能引起人们的关注并与之产生互动。

通过调查当前 Twitter 的流行度,哈伊(Hay,2010)研究了如何利用 Twitter 进行旅游营销,并提出两点值得思考的问题:人们能在 Twitter 或其他社交媒体上与一个组织或一个目的地成为好友吗?

当一个组织或一个账号并不代表一个人时,谁会看它的推文?

## 在 YouTube 上做广告

由于大多数人成长于通过视觉影像来学习的教育背景,因此社交媒体广告商往往会利用图像和视频平台进行市场营销,如 YouTube 和 Pinterest。YouTube 比电视广告具有更多的优势。首先,这是一个免费平台,大多数人都会在寻找特定内容时使用它。其次,与电视不同的是,你可以在任何时间、任何地点无限次地观看 YouTube 视频。

在 YouTube 上进行营销既简单又直观,并可以通过 Google AdWords 进行管理。企业可以上传他们想要展示的视频并对活动开销做预算;可以根据人口特征、主题、兴趣和关键字来选择目标受众。通过 YouTube 账号中的谷歌分析标签,广告商可以进一步了解他们的受众群体;通过 AdWords 账户,他们可以追踪受众的观看量、点击量和预算详情。其他有效的策略包括让网站所有者在其网站上插入视频,以及在其他社交媒体网站上分享 YouTube 视频。尽管观众可以选择跳过广告,但 YouTube 上的广告仍然是非常有效的。

## 在 Pinterest 上做广告

2014 年,另一个图像分享社交网站 Pinterest 宣布推出付费广告。像 Dillards 这类零售商的主页上出现了"Pin It"按钮,允许用户从网站上保存图片。付费广告即是新事物。那些付费广告也被叫做推广信息(Promoted Pins)。不同于普通的信息,推广信息在底部附上了免责声明,并在图像周围设置了红色边框,以使其独具一格。在笔者写这本书的时候,Pinterest 正准备推出广告服务。

其广告费用比其他社交媒体要高,平均每增加一次广告曝光就要花费30美元(McDermott,2014)。目前,该推广价对于本地商家来说可能过于昂贵,因此主要会是一些较大的品牌通过Pinterest发布广告。相关广告的主要受众群体是年轻女性,这与Pinterest的用户特征是一致的。而根据pinterest.com网站的统计,平均每个信息能够得到11次转发。

## 在LinkedIn上做广告

与其他社交媒体类似,LinkedIn允许企业用多种方式进行广告推广,包括在Linkedin feed中的直接赞助内容和赞助更新。公司可以制作一个包含文字、图片或视频的广告,也可以使用现有的或是新的内容与LinkedIn的用户一起分享。企业可以通过职称、工作职能、行业、地理位置、公司名称和规模大小来选择目标受众。根据公司的预算,他们可以选择按照点击量付费(每次点击费用,Cost per click, CPC),或者按照在网站页面中显示的次数支付费用(按每1 000次显示为单位计算费用)。目前,按点击量付费的广告最低价是一次点击2美元。广告商还可以选择广告投放时间的长短。在LinkedIn上投放广告的优势之一在于,广告商有机会能够准确地定位到受众群体。例如,一个正在寻求更多生源的法学院,可以只针对LinkedIn中职称为"法律助理"的人群投放广告。由此,他们不仅可以按照地理位置定位这些职业人群,也可以知道具备某一特定职称的人数。

## 病毒营销

病毒营销或互联网口碑营销,是一种"无偿的、通过同侪之间口耳相传某组织广告内容的传播模式,它通过网络来说服或影响受众将广告信息传递给他人"(Porter & Golan, 2006, p.33)。病毒营销起源于电子邮件设置,随着社交网站的普及而流行起来。研究发现,当发件人将信息传递给朋友或粉丝时,意在体验积极的情感(Phelps, Lewis, Mobillo, Perry, & Raman, 2004)。最受欢迎的病毒营销是视频类型的。这些广告通常是在电视上首播的视频片段,如超级碗广告(Dafonte-Gomez, 2013)。针对成功的病毒营销视频广告的分析显示,此类广告大多包含"惊喜"元素(Dafonte-Gomez, 2013)。埃克勒和博尔斯(Eckler & Bolls, 2011)也指出,为了能够让观众主动分享和传播视频,广告必须激发受众产生某种情绪。在用户传播方面,最成功的视频广告是包含关于性和裸露相关内容的广告;同时,由于它们不受美国联邦通信委员会(Federal Communications Commission)规范的约束,因此整体来说它们比电视广告尺度更大(Porter & Golan, 2006)。就视频长度而言,达冯特-戈麦斯(Dafonte-Gomez, 2013)发现病毒视频广告与标准电视广告没有明显区别,但是它们更有效率,因为消费者的信息来源于更可信的地方(他们的朋友)而非广告商(Nyilasy, 2004)。

纳拉亚南等人(Narayanan et al., 2012)认为,网络营销的理念是"通过推广活动来定位到某一特定人群,并让他们通过人际网络传播信息"。以 Twitter 为例,通常这些推广活动的重点对象是那些能为他们带来粉丝的人。一个人在社交网络中人气越高,他在社群中的影响力就越大。这是病毒营销的关键因素,不同的营销活动会针对特定的目标用户所在的社群进行推广。

# 总　　结

总体而言,企业在社交媒体上采取的策略应与企业的市场营销目标和目标受众人群保持一致。企业不仅需要关注自己发布的内容,也要分析人们在社交媒体上通常会关注什么内容、追随什么潮流趋势以及能被什么内容吸引。因此,与在其他类型的媒体上做的一样,企业在社交媒体平台上的市场营销策略也应有规划。

社交媒体广告为企业提供了一个与现有和潜在消费者进行互动的免费平台,使得企业能更好地理解用户的产品需求,并提供一些特价产品。随着人们对社交媒体娱乐与信息依赖度的加大,我们可以期待看到社交媒体广告的进一步发展。这不仅能使消费者与广告商建立更多的个人联系,还能让其在虚拟空间中与朋友和他人就产品与服务进行讨论。

# 参 考 文 献

Bucy, E. P. (2003). The interactivity paradox: Closer to the news but confused. In E. P. Bucy & J. E. Newhagen (Eds.), *Media access: Social and psychological dimensions of new technology use* (pp. 47-72). Mahwah, NJ: Erlbaum.

Corstjens, M., & Umblijs, A. (2012). The power of evil: The damage of negative social media strongly outweigh positive contributions. *Journal of Advertising Research*, 52(4), 433-449. doi: 10.2501/JAR-52-4-433-449.

Dafonte-Gomez (2013). The key elements of viral advertising. From motivation to emotion in the most shared videos. *Comunicar*, 22(43), 199-206.

Eckler, P., & Bolls, P. (2011). Spreading the virus: Emotional tone of viral advertising and its effect on forwarding intention and attitudes. *Journal of Interactive Advertising*, 11, 1-11. doi: 10.1080/15252019.2011.10722180.

Emanuelli, E. (2012). *How to increase interactivity on your Facebook fan page*. Retrieved from http://onlineincometeacher.com/socialmedia/increase-interactivity-on-facebook-fan-page/.

Fogg, B. J. (2003). *Persuasive technology: Using computers to change what we think and do*. Boston: Morgan Kaufmann.

Gommans, M., Krishnan, K.S. & Scheffold, K. B. (2001). From brand loyalty to e-loyalty: A conceptual framework. *Journal of Economic and Social Research*, *3*, 43-58. doi: 10.1.1.105.3103.

Habibi, M. R., Laroche, M., & Richard, M. -O. (2014). The roles of brand community and community engagement in building brand trust on social media. *Computers in Human Behavior*, *37*, 152-161. doi: 10.1016/j.chb.2014.04.016.

Hay, B. (2010). *Twitter Twitter-but who is listening? A review of the current and potential use of twittering as a tourism marketing tool*. Presented at the CAUTHE 2010 20th International Research Conference.

Hill, R. P., & Moran, N. (2011). Social marketing meets interactive media: Lessons for the advertising community. *International Journal of Advertising*, *30*(5), 815-838. doi: 10.2501/IJA-30-5-815-838.

Kiousis, S. (2002). Interactivity: A concept explication. *New Media & Society*, *4*(3), 355-383. doi: 10.1177/146144402320564392.

Manjoo, F. (2013). *Facebook followed you to the supermarket*. Retrieved from http://www.slate.com/articles/technology/technology/2013/03/facebook_advertisement_studies_their_ads_are_more_like_tv_ads_than_google.2.html.

McDermott, J. (2014). *How Pinterest is selling ads to agencies*. Retrieved from http://digiday.com/platforms/heres-pinterest-pitching-agencies-ads/.

Megna, M. (2009). *Facebook, Twitter, and social media marketing*. Retrieved from http://www.internetnews.com/ecnews/article.php/3839521/Facebookpercent2BTwitterpercent2Bandpercent2BSocialpercent2BMediapercent2BMarketing.htm.

Muniz, A. M., & O'Guinn, T. C. (2001). Brand community. *Journal of Consumer Research*, *27*(4), 412-432. doi: 10.1086/319618.

Narayanan, M., Asur, S., Nair, A., Rao, S., Kaushik, A., Mehta, D., Athalye, S. & Lalwani, R. (2012). Social media and business. *Vikalpa: The Journal for Decision Makers*, *37*(4), 69-111.

Nyilasy, G. (2004). *Word-of-mouth advertising: A 50-year review and two theoretical models for an online chatting context*. Paper presented at the

2004 Convention of the Association for Education in Journalism and Mass Communication, Toronto, Canada.

Peppers, D., Rogers, M. & Dorf, R. (1999). Is your company ready for one-to-one marketing. *Harvard Business Review*, 77, 151-60.

Phelps, J. E., Lewis, R., Mobillo, L., Perry, D., & Raman, N. (2004). Viral marketing or electronic word-of-mouth advertising: Examining consumer responses and motivations to pass along email. *Journal of Advertising Research*, 44, 333-48. doi: 10.1017/S0021849904040371.

Porter, L., & Golan, G. (2006). From subservient chickens to brawny men: A comparison of viral advertising to television advertising. *Journal of Interactive Advertising*, 6. Retrieved from http://www.jiad.org/article78.html.

Powers, T., Advincula, D., Austin, M. S., & Graiko, S. (2013). Digital and social media in the purchase-decision process: A special report from the Advertising Research Foundation. *Journal of Advertising Research*, 52, 479-489. doi: 10.2501/JAR-52-4-479-489.

Stelzner, M. (2014). *2014 social media marketing industry report*. Retrieved from http://www.socialmediaexaminer.com/social-media-marketing-industry-report-2014/.

Tucker, C. E. (2014). Social networks, personalized advertising, and privacy controls. *Journal of Marketing Research*, 51, 546-562. doi: 10.1509/jmr.10.0355.

Weinberg, T. (2009). *The new community rules: Marketing on the social web*. Sebastopol, CA: O'Reilly Media Inc.

Witkemper, C., Choong Hoon, L., & Waldburger, A. (2012). Social media and sports marketing: Examining the motivations and constraints of Twitter users. *Sport Marketing Quarterly*, 21(3), 170-183.

# 9

# 社交媒体成瘾

本书的前几章已经详细讨论了社交媒体的不足之处,如隐私和安全、网络欺诈和错误信息、自恋及人际关系的逐渐恶化等。本章将聚焦社交媒体成瘾,讨论社交媒体成瘾的定义、成因和后果。我们发现,尽管许多报纸文章都已经注意到社交媒体的负面效果,但相关研究仍然十分有限。

## 社交媒体成瘾定义的问题

尽管社交媒体网站有许多积极的用处,但仍存在着不足,过度使用社交媒体就是其中之一。尽管对社交媒体成瘾的定义很多,但明确的定义还没有。根据格里菲思(Griffiths, 2000)的定义,如果上网行为已经成为一个人生活中最重要的部分,那么我们可以认为他对互联网产生了过度依赖。该定义也包含另一层意思,即对互联网过度依赖的人,会在无法使用互联网时产生戒断症状(withdrawal symptoms)或消极情绪。其他一些可能出现的症状还包括情绪的变化,为了重新获得积极情绪而花费更多的时间上网或玩一个新游戏(Cash, Rae, Steel, & Winkler, 2012),减少使用互联网尝试的失败,出现烦躁不安和易怒的精神状态(Beard, 2005)。

社交媒体成瘾目前尚无统一的官方定义。大多数研究只是着眼于网络成瘾，因此出现了不同的概念被用来解释同一现象：有问题的互联网使用（Davis，2001）、互联网依赖（Dowling & Quirk，2009）、强迫性互联网使用、病理性互联网使用（Caplan，2002）和网络成瘾综合症。目前接受度最高的定义是由卞等人（Byun et al.，2009）提出的网络成瘾综合症（Internet Addiction Disorder），意指干扰到日常生活的过度的电脑使用。尽管该现象尚未被归为精神障碍，但是许多研究人员（如Cash et al.，2012）已经敦促相关组织考虑将其列入《精神障碍诊断与统计手册》（*Diagnostic and Statistical Manual of Mental Disorders*）。

# 社交媒体成瘾的成因

霍梅斯、卡恩斯和阿利克斯·蒂姆科（Hormes，Kearns，& Alix Timko，2014）解释了人们可能会沉迷于社交网站的原因。首先，这些社交网站存在一些能鼓励用户反复查阅网站的功能。例如，会有新的内容持续在网上发布；或是当有人评论我们的帖子时，我们会收到手机通知。这会导致我们不断重复刷新页面进行查阅。霍梅斯等人（2014）提出的另一个原因是，有些人情绪调节困难、对冲动的控制力差，且较难从事目标导向的行为。所以对某些用户来说，社交媒体成为一种逃避现实的方式，而另一些人则利用互联网来解压。总之，社交媒体内容刺激性的特质能让使用者达到心理兴奋，这也是许多人将有问题的互联网使用行为与其他类型的问题行为进行比较的原因，如赌博（Young & Nabuco，2011）或色情内容（Cash et al.，2012）。

有学者尝试用生物学概念对社交媒体成瘾进行解释。根据比尔德(Beard,2005)的研究,血清素/多巴胺含量不足的人更容易上瘾。血清素是一种负责维持情绪平衡的化学成分(medicinenet.com,2014)。而多巴胺则是一种控制兴奋、刺激和奖励的化学物质。它会因为受到奖励的体验(如食物和性)而在大脑中释放(Arias-Carrión & Pöppel, 2007)。在社交媒体上玩游戏或在Facebook上与他人聊天都可以获得所谓的奖励,从而导致多巴胺的增加(Cash et al., 2012)。

唐、维塔克和拉罗斯(Tong, Vitak, & Larose, 2010)也认为,事实上,许多社交媒体的负面影响(如成瘾)都源自用户的消极使用方式。用户本身可能就有一些与社交媒体相关联的消极面。因此社交媒体带来的负面影响并不意味着社交媒体本身就是消极的,而是由于那些使用它的人导致的。例如,有研究人员(Aboujaoude, Koran, Gamel, Large, & Serpe, 2006)发现,每八个美国人中,就会有至少一个人在互联网使用行为上可能存在问题。当观察3D《第二人生》(Second Life)虚拟世界中的用户时,吉尔伯特、墨菲和麦克纳利(Gilbert, Murphy, & McNally, 2011)发现,大约有三分之一的玩家符合网络成瘾的标准。此外,他们的成瘾还与一些现实生活中的问题行为有关,如购物、性、赌博、药物和饮酒成瘾等。这表明人格特质在一定程度上可以预示社交媒体成瘾。事实上,多项研究发现都指出,自尊心低、害羞、内向、神经质、高度孤独感和抑郁症都可能会导致个人在网络上耗费太多时间(Cao & Su, 2006; Griffiths & Dancaster, 1995; Smahel, Brown, & Blinka, 2012)。还有研究认为抑郁症和Facebook成瘾之间呈现正相关(Hong, Huang, Lin, & Chiu, 2014)。据洪等人(Hong et al.)表示,当一个人无法控制自己在Facebook上的行为时,说明他/她患上了Facebook成瘾综合症。

有研究表明,女大学生对Facebook的依赖整体上比男大学生更明显(Hormes et al., 2014)。这可能是由于女性已经在Facebook上花了更多的时间(Sheldon, 2008),因为她们更关注人际关系的维持,这就是她们使用Facebook的主要原因。其他研究表明,自我导向性低的人在网络成瘾方面得分较高,因为他们无法应对日常生活中出现的问题(Kose, 2003; LaRose, Lin, & Eastin, 2003)。蒙塔格等人(Montag et al., 2011)指出,让这些上瘾者感受到成就感可以提高他们的自尊心。此外,那些具备高度自觉意识的人较少产生有问题的互联网使用行为(Montag et al., 2011)。这些人通常在工作的时候有很强的责任感且办事效率高。

大量文献综述表明,互联网上瘾者占总人口的0.3%—10.6%(Shaw & Black, 2008),而过度使用行为在年轻人中最为常见。在一项针对10个不同国家的12个校区的研究中(Moeller, 2010),研究者发现了类似于社交媒体成瘾的症状。当年轻人(17—23岁)被要求24小时内不得使用社交媒体时,他们出现了类似毒瘾的症状。80%的学生出现精神上和身体上的困扰、恐慌、混乱。很多人感到孤独、沮丧、悲伤和无聊。根据莫雷诺等人(Moreno et al., 2011)的研究,最常与社交网站使用相关联的精神疾病是抑郁症。

一些研究者认为可能存在关于社交媒体成瘾的化学解释。霍恩(Horn, 2012)指出,用户在登入社交网站时,会出现心理生理反应,这类似于用户在从事创意性活动时的反应。用户只需登入Facebook这类网站就可以分泌肾上腺素。"赞"这一状态可以让用户感到获得了奖励,从而增加了多巴胺的分泌。格罗斯曼(Grossman, 2007)将这一成瘾与过量毒品成瘾相比较,如可卡因或快克。其他研究人员(如Alavi et al., 2012)则发现行为成瘾如网络成瘾,可能与物质成瘾相类似。例如,在上述针对10个不同国家的12个校区的研究中,有参与者提出,感觉"像瘾君子一样痒",感到麻痹、有压力及焦虑(Moeller,2010)。社交媒体的另一个问题在于,相比其他昂贵的物质来说,它们免费并且易获得。

虽然物质的上瘾可以通过完全节制的方法来进行控制,但对我们所生存的社会来说,要求年轻人放弃使用社交媒体几乎是不可能的。因此,有学者建议可以控制人们使用社交媒体的时间(Griffiths,2013)。霍梅斯等人(2014)提出治疗技术需要与治疗进食障碍相类似,包括使之正常化,而不是完全消除问题行为。其他治疗网络成瘾的建议还包括通过外部机制来抑制社交媒体成瘾,如设置自我提醒小卡片、加入帮助小组等(Young,1999)。埃德尔斯坦(Edelstein,2014)还建议人们在周末尝试离开社交媒体,可以设置一些使用规则,有目的性地查看(如只看亲友的婚礼照片),以及设定闹钟。

但是目前仍没有足够的研究能够证明这些技术是否可以实际运用于治疗社交媒体成瘾。

## 社交媒体成瘾的负面后果

互联网和社交媒体的上瘾还可能会带来一系列负面后果,包括不良的学校表现(Tsitsika, Critselis, Louizou, Janikian, Freskou, Marangou, et al., 2011)、较低的工作效率和低质量的人际关系(Milani, Osualdella, & Di Blasio, 2009)。青少年会经常查看其他人发布在Facebook或Twitter上的内容。他们认为好友和粉丝的数量、获得赞的数量都比友谊的质量更加重要。因为非言语暗示是不需要的,在Facebook上说"生日快乐"比通过打电话或面对面说更加简单。但是,人际关系质量的下降并不是由社交媒体引发的。在1998年,一批研究人员就发现,那些在网络上花费时间越多的人,越少与家人进行沟通(Kraut, Patterson, Lundmark, Kiesler, Mukophadhyay, & Scherlis)。到目前为止,我们仍然没有足够的研究去解释过去十年来人际关系的质量是如何变化的。但是当我们与朋友在餐厅用餐时,大家突然变得更着迷于手机,而不是交谈。根据最新的皮尤研究互联网项目(Pew Research Internet Project, 2014),截至2014年1月,有90%的美国人拥有手机,其中近60%是智能手机。

总之,人们普遍认为,任何形式的成瘾都是由生物、社会和心理因素综合造成的(Griffiths, 2005)。当论及社交技巧时,喜欢在线上沟通而非线下沟通的人的自我表达能力往往会不足(Griffiths, 2013)。他们沉迷于社交媒体,尤其是社交网站,因为他们可以在线上通过任何方式进行自我表达。在短期内,这会产生更高的生活满意度;然而,从长远来看,却会对工作和学习产生负面的后果。

# 参 考 文 献

Aboujaoude, E., Koran, L. M., Gamel, N., Large, M., & Serpe, R. (2006). Potential markers for problematic Internet use: A telephone survey of 2,513 adults. *CNS Spectrums*, 11(10), 750-755.

Alavi, S., Ferdosi, M., Jannatifard, F., Eslami, M., Alaghemandan, H., & Setare, M. (2012). Behavioral addiction versus substance addiction: correspondence of psychiatric and psychological views. *International Journal of Preventative Medicine*, 3, 290-294.

Arias-Carrión, Ó., & Pöppel, E. (2007). Dopamine, learning, and reward-seeking behavior. *Acta Neurobiologiae Experimentalis*, 67(4), 481-488.

Beard, K. W. (2005). Internet addiction: A review of current assessment techniques and potential assessment questions. *CyberPsychology & Behavior*, *8*, 7-14. doi: 10.1089/cpb.2005.8.7.

Byun, S., Ruffini, C., Mills, J., Douglas, A., Niang, M., Stepchenkova, S., Lee, S., Loutfi, J., Lee, J., Atallah, M., & Blanton, M. (2009). Internet addiction: Metasynthesis of 1996 - 2006 quantitative research. *CyberPsychology & Behavior*, *12*(2), 203 - 207. doi: 10.1089/cpb.2008.0102.

Cash, H., Rae, C. D., Steel, A. H., & Winkler, A. (2012). Internet addiction: A brief summary of research and practice. *Current Psychiatry Reviews*, *8*(4), 292-298. doi: 10.2174/157340012803520513.

Cao, F., & Su, L. (2006). Internet addiction among Chinese adolescents: prevalence and psychological features. *Child: Care, Health & Development*, *33*(3), 275-281.

Caplan, S. E. (2002). Problematic Internet use and psychosocial well-being: development of a theory-based cognitive-behavioral measurement instrument. *Computers in Human Behavior*, *18*, 553-575. doi: 10.1016/S0747-5632(02)00004-3.

Davis, R. A. (2001). A cognitive behavioral model of pathological internet use (PIU). *Computers in Human Behavior*, *17*, 187-195. doi: 10.1016/S0747-5632(00)00041-8.

Dowling, N. A, & Quirk, K. L. (2009). Screening for Internet dependence: Do the proposed diagnostic criteria differentiate normal from dependent Internet use? *CyberPsychology & Behavior*, *12*, 21 - 27. doi: 10.1089/cpb.2008.0162.

Edelstein, J. (2014). Break free from social media. *Real Simple*, *15*(1), 28.

Gilbert, R., Murphy, N., & McNally, T. (2011). Addiction to the 3 - dimensional Internet: estimated prevalence and relationship to real world addictions. *Addiction Research & Theory*, *19*, 380-390. doi: 10.3109/16066359.2010.530714.

Griffiths, M. (2000). Does Internet and computer "addiction" exist? Some case study evidence. *CyberPsychology & Behavior*, *3*, 211-218. doi: 10.1089/109493100316067.

Griffiths, M. (2005). A "components" model of addiction within a biopsychosocial framework. *Journal of Substance Use*, *10*, 191-197. doi: 10.1080/14659890500114359.

Griffiths, M. (2013). Social networking addiction: Emerging themes and issues. *Addiction: Research & Therapy*, *4*. doi: 10.4172/2155-6105.1000e11. Retrieved from http://omicsonline.org/social-networking-addiction-emerging-themes-and-issues-2155-6105.1000e118.pdf.

Griffiths, M. D., & Dancaster, I. (1995). The effect of type A personality on physiological arousal while playing computer games. *Addictive Behaviors*, *20*, 543-548. doi: 10.1016/0306-4603(95)00001-S.

Grossman, L. (2007). The hyperconnected. *Time*, *169*(16), 54-56.

Hong, F., Huang, D., Lin, H., & Chiu, S. (2014). Analysis of the psychological traits, Facebook usage, and Facebook addiction model of Taiwanese university students. *Telematics and Informatics*, *31*, 597-606. doi: 10.1016/j.tele.2014.01.001.

Hormes, J. M., Kearns, B., & Alix Timko, C. (2014). Craving Facebook? Behavioral addiction to online social networking and its association with emotion regulation deficits. *Addiction*, *109*, 2079-2088. doi: 10.1111/add.12713.

Horn, L. (2012). Study finds chemical reason behind Facebook 'addiction.' *PC Magazine*, 1.

Kose, S. (2003). Psychobiological model of temperament and character. *Yeni Symposium*, *41*, 86-97.

Kraut, R., Patterson, M., Lundmark, V., Kiesler, S., Mukopadhyay, T., & Scherlis, W. (1998). Internet paradox: A social technology that reduces social involvement and psychological well-being? *American Psychologist*, *53*, 1017-1031. doi: 10.1037/0003-066X.53.9.1017.

LaRose, R., Lin, C. A., & Eastin, M. (2003). Unregulated Internet usage: Addiction, habit, or deficient self-regulation? *Media Psychology*, *5*, 225-253. doi: 10.1207/S1532785XMEP0503_01.

Medicinenet.com (2014). *Definition of serotonin*. Retrieved from http://www.medicinenet.com/script/main/art.asp?articlekey=5468.

Milani, L., Osualdella, D., & Di Blasio, P. (2009). Quality of interpersonal relationships and problematic Internet use in adolescence. *CyberPsychology &*

*Behavior*, *12*(6), 681-684. doi: 10.1089/cpb.2009.0071.

Moeller, S. (2010). *The world unplugged*. Retrieved from http://theworldunplugged.wordpress.com/.

Montag, C., Flierl, M., Markett, S., Walter, N., Jurkiewicz, M., & Reuter, M. (2011). Internet addiction and personality in first-person-shooter video gamers. *Journal of Media Psychology*, *23*, 163-173. doi: 10.1027/1864-1105/a000049.

Moreno, M., Jelenchick, L., Egan, K., Cox, E., Young, H., Gannon, K., & Becker, T. (2011). Feeling bad on Facebook: Depression disclosures by college students on a social networking site. *Depression and Anxiety*, *28*(6), 447-455. doi: 10.1002/da.20805.

Pew Research Internet Project (2014). *Mobile technology fact sheet*. Retrieved from http://www.pewinternet.org/fact-sheets/mobile-technology-fact-sheet/.

Shaw, M., & Black, D. W. (2008). Internet addiction: Definition, assessment, epidemiology and clinical management. *CNS Drugs*, *22*, 353-365.

Sheldon, P. (2008). Student favorite: Facebook & motives for its use. *Southwestern Mass Communication Journal*, *23*, 39-55.

Smahel, D., Brown, B. B., and Blinka, L. (2012). Associations between online friendship and Internet addiction among adolescents and emerging adults. *Developmental Psychology*, *48*, 381-388. doi: 10.1037/a0027025.

Tong, S., Vitak, J., & LaRose, R. (2010). Truly problematic or merely habitual? An integrated model of the negative consequences of social networking. *Conference Papers—International Communication Association*.

Tsitsika, A., Critselis, E., Louizou, A., Janikian, M., Freskou, A., Marangou, E., et al. (2011). Determinants of Internet addiction among adolescents: A case-control study. *The Scientific World Journal*, *11*, 866-874. doi: 10.1100/tsw.2011.85.

Young, K. S. (1999). Internet addiction: Symptoms, evaluation, and treatment. *Innovations in clinical practice*. Retrieved from http://www.netaddiction.com/articles/symptoms.pdf.

Young, K. S, & Nabuco, de Abreu C. (2011). *Internet addiction: A handbook and guide to evaluation and treatment*. New Jersey: John Wiley & Sons Inc.

附录

# 社交媒体简史

社交媒体的历史是从1978年第一个虚拟在线社区的创建开始的。当时计算机科学家默里·图罗夫(Murray Turoff)和罗克珊·希尔茨(S. Roxanne Hiltz)在美国新泽西理工学院建立了电子信息交换系统(electronic information exchange system, EIES)(Acar, 2008),被视为首批集体智慧项目之一。EIES能够让用户发送电子邮件,查看公告板并使用列表服务(Hiltz & Turoff, 1978; 1993; Wasserman & Faust, 1994)。该系统还被用于教授课程、召开会议和促进研究。EIES项目由国家科学基金会资助,用于进一步探索计算机会议的潜力。该项目拥有2 000多名来自Exxon、IBM、政府机构和美国大学的用户("IRC History", 2000)。在EIES项目完成的同年,1978年,"电子公告栏系统"(bulletin board system)被发明出来。第一个公告栏基于文本,能让两台或多台计算机之间使用调制解调器(modem)和电话线进行信息传递。直到20世纪90年代初,在万维网出现之前,它们一直都是最主流的在线社区("The BBC Corner", 2009)。1989年,蒂姆·伯纳斯-李(Tim Berners-Lee)创立了万维网(World Wide Web),不久之后,便开启了博客时代。

第一篇博客(blog)是1994年由美国史瓦兹摩尔学院(Swarthmore College)的学生贾斯汀·霍尔(Justin Hall)撰写的发表在www.links.net上的个人日记《秘密的贾斯汀的链接》

(*Justin's Links from the Underground*),当时他是旧金山《连线》(*Wired*)杂志的实习生(Harmanci,2005)。在1997年,Robot Wisdom(robotwisdom.com)的约翰·巴杰(Jorn Barger)创造了术语"weblog"一词,意思是记录网络。后来,彼得·摩霍兹(Peter Merholz)拆分了这个词变成"we blog",并缩短为"blog"(博客)。Pyra Labs的埃文·威廉斯(Evan Williams)将"blog"同时作为动词与名词使用,并设计了一个新的术语"blogger"(博主)("Origins of Blog",2008)。

1998—1999年,陆续出现了一批最受欢迎的博客平台。Open Diary于1998年推出,LiveJournal和Blogger在1999年创立。2000年,安德鲁·苏利万(Andrew Sullivan)推出了Daily Dish,这是最早的政治博客之一,2001年就涌现出大量的政治博客。有一些人因博客而成名。2002年,希瑟·阿姆斯特朗(Heather Armstrong)因为在她的个人博客dooce.com中记录职场上遇到的人而被解雇。术语"dooced"由此出现,意思是"由于个人网站而失业"(urbandictionary.com)。在2003—2004年社交网站出现之前,博客是最受欢迎的社交媒体。

第一个在线社交网络于1997年创建,名字是SixDegrees.com(boyd & Ellison,2007)。创始人安德鲁·魏因赖希(Andrew Weinreich)的这个想法来源于米尔格拉姆(Milgram)著名的小世界研究。这一研究提出,通过六个或更少的人的关系,就可以将任何两个人联系在一起。阿德里安·斯科特(Adrian Scott)在2001年创建了一个名为Ryze.com的社交网站,旨在连接职场专业人士,这一想法与LinkedIn很相似。该网站的名称来源于"rise up"(提升),"因为这关乎用户通过有质量的网络相互帮助提升彼此"。2002年,加拿大的程序员乔纳森·艾布拉姆斯(Jonathan

Abrams)推出了 Friendster.com,旨在为朋友之间的线下聚会创造条件。这是第一个达到百万会员的网站,但在 MySpace 和 Facebook 出现后开始流失用户,后来被重新设计为本地游戏网站(boyd & Ellison, 2007;"Friendster Back", 2012)。2003 年 8 月,克里斯·德沃尔夫和汤姆·安德森(Chris deWolfe & Tom Anderson)创建了 Myspace.com(Lapinski, 2006),这是一个流行音乐的聚集地,很多音乐家的职业生涯从此处开始,并且自拍的想法就来源于 MySpace。2005—2008 年,MySpace 都是世界上访问量最多的网站,在 2006 年超过了 Google(Cashmore, 2006)。2008 年 4 月,Facebook 的访问量超过了 MySpace,此后 MySpace 的访问量开始逐渐下降(Albanesius, 2009)。在 2003 年、2004 年,出现了许多流行的社交网站。LinkedIn 是一个面向职场的社交网站,于 2003 年在美国加利福尼亚的山景城创建,创始人为里德·霍夫曼、艾伦·布卢、康斯坦丁·居里克、埃里克·利和让鲁克·瓦扬(Reid Hoffman, Allen Blue, Konstantin Guericke, Eric Ly, & Jean-Luc Vaillant)。2004 年,Facebook 成为 MySpace 的主要竞争对手。马克·扎克伯格(Mark Zuckerberg)创建了 facebook.com,以取代纸质的课程目录,作为对大学新生介绍学校信息的一部分(Carlson, 2012)。该网站最初的成员仅限于哈佛大学的学生,随后扩大到波士顿地区和常春藤联盟的其他高校。之后继续扩大到所有大学生、高中生,到最后,凡是 13 岁以上的人都能成为 Facebook 用户。2006 年,杰克·多尔塞、诺厄·格拉斯、比扎·斯通、埃文·威廉斯(Jack Dorsey, Noah Glass, Biz Stone, & Evan Williams)共同创建了另一个非常受欢迎的社交网站——Twitter。Twitter 可以用来发送和接收短消息(不超过 140 个字符)。2010 年,Instagram 由凯文·瑟斯特朗姆和麦克·克里格(Kevin

Systrom & Mike Krieger)创立。另一个受欢迎的图像共享社交网站 Pinterest 也于同年创建(Mull & Lee,2014)。2012 年,Google 推出了 Google+。

维基(wiki)是另一种形式的社交媒体。"维基"一词在夏威夷语中是"快速"的意思。维基是一个协作项目,被视为社交媒体应用的一种特殊形式,因为它可以实现多用户联合创建知识性内容(Kaplan & Haenlein,2014)。维基是最民主的社交媒体形式,因为它允许任何人使用简单的浏览器来添加、修改或删除网页上的内容(Kaplan & Haenlein,2014)。目前最受欢迎的维基是维基百科(Wikipedia),它是由吉米·威尔士和拉里·桑格(Jimmy Wales & Larry Sanger)于 2001 年创立的(Sidener,2004),是可以免费访问的多语言在线百科全书。维基百科改变了知识储存与访问的方式。不同于只有专家才能发布词条的传统媒体,在维基百科上任何读者都可以成为一名编辑。

2005 年,三位 PayPal 的员工史蒂夫·陈、查德·赫尔利和贾韦德·卡里姆(Steve Chen, Chad Hurley, & Jawed Karim)共同创建了视频分享网站 YouTube。该网站能够让用户上传自己的视频并观看其他用户的视频。YouTube 目前已经被不同的受众所使用,包括私人用户、在网站上发布广告的企业、政治家和政府。YouTube 在 2010 年的阿拉伯之春运动中发挥了非常重要的作用。有些人也因为他们在 YouTube 上的表现而变得有名,其中之一是苏珊·博伊尔(Susan Boyle),一位因参加《英国达人》而成名的苏格兰歌手。

# 参 考 文 献

Acar, A. (2008). Antecedents and consequences of online social networking behavior: The case of Facebook. *Journal of Website Promotion*, *3*, 62-83. doi: 10.1080/15533610802052654.

Albanesius, C. (2009, June 16). *More Americans go to Facebook than MySpace*. Retrieved from http://www. pcmag. com /article2 /0, 2817, 2348822,00.asp.

"The BBC Corner—A Brief History of BBS Systems" (2009). Retrieved from http://www.bbscorner.com/usersinfo/bbshistory.htm.

boyd, d. m., & Ellison, N. B. (2007). Social network sites: Definition, history, and scholarship. *Journal of Computer-Mediated Communication*, *13*(1), article 11. Retrieved from http://jcmc.indiana.edu/vol13/issue1/Boyd.ellison.html.

Carlson, N. (2012, May 1). Inside Pinterest: An overnight success four years in the making. *Business Insider*. Retrieved from http://www.businessinsider.com/inside-pinterest-an-overnight-success-four-years-in-the-making-2012-4.

Cashmore, P. (2006, July 11). *MySpace, America's number one*. Retrieved from http://mashable.com/2006/07/11/myspace-americas-number-one/.

"Friendster back with social network" (2012). *Social media today*. Retrieved from http://socialmediatoday.com/mohammed-anzil/1113891/friendster-back-social-network.

Harmanci, R. (2005, February 20). Time to get a life—pioneer blogger Justin Hall bows out at 31. *San Francisco Chronicle*. Retrieved from http://www.sfgate.com/newsarticle/Time-to-get-a-life-pioneer-blogger-Justin-Hall-2697359.php.

Hiltz, S. R., & Turoff, M. (1978). *The network nation: Human communication via computer*. Reading, MA: Addison-Wesley.

Hiltz, S. R., & Turoff, M. (1993). *The network nation: Human communication via computer, revised edition*. Cambridge, MA: MIT Press.

"IRC history—Electronic Information Exchange System" (2000). Retrieved from http://www.livinginternet.com/r/ri_eies.htm.

Kaplan, A., & Haenlein, M. (2014). Collaborative projects (social media application): About Wikipedia, the free encyclopedia. *Business Horizons*, *57*(5), 617-626. doi: 10.1016/j.bushor.2014.05.004.

Lapinski, T. (2006). MySpace: The business of spam 2. 0. ValleyWag.

Retrieved from http://valleywag.com/tech/myspace/myspace-the-business-of-spam-20-exhaustive-edition-199924.php.

Mull, I. R., & Lee, S. (2014). "PIN" pointing the motivational dimensions behind Pinterest. *Computers in Human Behavior*, *33*, 192–200. doi: 10.1016/j.chb.2014.01.011.

Origins of "Blog" and "Blogger." American Dialect Society Mailing List (2008, April 20).

Sidener, J. (2004, December 6). *Everyone's encyclopedia*. Retrieved from http://www.utsandiego.com/uniontrib/20041206/news_mz1b6encyclo.html.

Wasserman, S., & Faust, K. (1994). *Social network analysis: Methods and applications*. Cambridge, UK: Cambridge University Press.

图书在版编目(CIP)数据

社交媒体:原理与应用/[美]帕维卡·谢尔顿著;张振维译.—上海:
复旦大学出版社,2018.4(2019.10重印)
(复旦新闻与传播学译库)
书名原文:Social Media:Principles and Applications
ISBN 978-7-309-13603-6

Ⅰ.社… Ⅱ.①帕…②张… Ⅲ.互联网络-传播媒介-研究 Ⅳ.G206.2

中国版本图书馆 CIP 数据核字(2018)第 058685 号

Copyright© 2015 by Lexington Books
All rights reserved.
Fudan University Press Co., Ltd. is authorized to publish and distribute exclusively the Chinese
Simplified language edition. This edition is authorized for sale throughout Mainland of China.
No part of the publication may be reproduced or distributed by any means, or stored in a
database or retrieval system, without written permission of the publisher.

上海市版权局著作权合同登记号:图字 09-2018-115

社交媒体:原理与应用
[美]帕维卡·谢尔顿 著  张振维 译
责任编辑/朱安奇
复旦大学出版社有限公司出版发行
上海市国权路 579 号  邮编:200433
网址:fupnet@fudanpress.com  http://www.fudanpress.com
门市零售:86-21-65642857  团体订购:86-21-65118853
外埠邮购:86-21-65109143  出版部电话: 86-21-65642845
常熟市华顺印刷有限公司

开本 890×1240  1/32  印张 6.125  字数 135 千
2019 年 10 月第 1 版第 2 次印刷

ISBN 978-7-309-13603-6/G·1826
定价:32.00 元

如有印装质量问题,请向复旦大学出版社有限公司出版部调换。
版权所有  侵权必究